4th International Conference on Information Processing and Control Engineering (ICIPCE 2020)

AA007526

IOP Conference Series: Materials Science and Engineering
Volume 946

Berkeley, California, USA
27 - 29 June 2020

ISBN: 978-1-7138-3168-6
ISSN: 1757-8981

Printed from e-media with permission by:

Curran Associates, Inc.
57 Morehouse Lane
Red Hook, NY 12571

CURRAN ASSOCIATES INC. proceedings.com

Some format issues inherent in the e-media version may also appear in this print version.

This work is licensed under a Creative Commons Attribution 3.0 International Licence.
Licence details: http://creativecommons.org/licenses/by/3.0/.

No changes have been made to the content of these proceedings. There may be changes to pagination and minor adjustments for aesthetics.

Printed with permission by Curran Associates, Inc. (2021)

For permission requests, please contact the Institute of Physics at the address below.

Institute of Physics
Dirac House, Temple Back
Bristol BS1 6BE UK

Phone: 44 1 17 929 7481
Fax: 44 1 17 920 0979

techtracking@iop.org

Additional copies of this publication are available from:

Curran Associates, Inc.
57 Morehouse Lane
Red Hook, NY 12571 USA
Phone: 845-758-0400
Fax: 845-758-2633
Email: curran@proceedings.com
Web: www.proceedings.com

TABLE OF CONTENTS

ADVANCED INFORMATION TECHNOLOGY AND APPLICATION

MODERN ELECTRONIC THEORY AND TECHNOLOGY

IOP Conf. Series: Materials Science and Engineering **946** (2020) 011001 doi:10.1088/1757-899X/946/1/011001

PREFACE

The 4th International Conference on Information Processing and Control Engineering (ICIPCE 2020) was held online at University of California at Berkeley, the United States during June 27-29, 2020.

The main topic of the conference is information processing and control, but certainly not limited to this. Information processing in the modern world is of great importance, which presents in so many areas of engineering and daily life. Thus, such a conference has scientific and applied value.

Nowadays, technologies related to information processing and management are developing rapidly. Russia has a good position in this area, so the choice of venue for such a conference is well justified.

At the conference, various reports were presented on the designated topics. There were representatives of universities, companies and academic institutions. The geography of participants included China, India, Kazakhstan, and so on. Invited keynote speakers gave a 45-minute speech, each author's presentation was finished within 15 minutes. Through zoom, the authors could share the screen and discuss the works actively. The conference came to a successful end with attendees' support.

The conference consisted of two sections: Advanced information technology and application, Modern electronic theory and technology.

This conference is an excellent opportunity for scientists from different countries to meet with each other and discuss common tasks in a particular area of their research. Scientific schools, which include various scientists from different countries, will receive excellent experience from the possibility of such communication among scientists.

Conference Committee Chair
Prof. Dennis K. Lieu
University of California, Berkeley, USA

Content from this work may be used under the terms of the Creative Commons Attribution 3.0 licence. Any further distribution of this work must maintain attribution to the author(s) and the title of the work, journal citation and DOI.
Published under licence by IOP Publishing Ltd

IOP Conf. Series: Materials Science and Engineering **946** (2020) 011002 doi:10.1088/1757-899X/946/1/011002

International Advisory Committee

Prof. Allison M. Okamura (IEEE Fellow), Stanford University, United States

Conference Committees Chair

Prof. Dennis Lieu, University of California, Berkeley, United States

Program Committee Chairs

Prof. Dr. Jan Awrejcewicz, Lodz University of Technology, Poland
Marina Barulina, Institute of Precision Mechanics and Control, Russian Academy of Sciences, Russia

Technical Committee Chairs

Prof. He Xu, Harbin Engineering University, China
Prof. Kunhong Liu, Xiamen University, China
Prof. Kwanho You, Sungkyunkwan University, South Korea
Prof. Mohamed bin Hussein, Universiti Technologi Malaysia, Malaysia
Assoc. Prof. Yi Jin, University of Science and Technology of China, China
Assoc. Prof. Jie Yang, Dalian University of Technology, China
Assoc. Prof. Dirman Hanafi, Universiti Tun Hussein Onn Malaysia, Malaysia
Assoc. Prof. Dr. Vyacheslav Petrenko,North-Caucasus Federal University, Russia
Assoc. Prof. Dr. Maria Lapina, North-Caucasus Federal University, Russia
Asst. Prof. Abdel-Hameed Badawy, New Mexico State University, USA
Dr. Zahid Akhtar, University of Memphis, USA
Dr.Hsien-I Lin, National Taipei University of Technology, Taiwan
Dr. Lihong Zheng, Charles Sturt University, Australia
Dr. Adrian Kampa, Silesian University of Technology, Poland
Sze-Teng Liong, Feng Chia University, Taiwan
Muath Alrammal, Higher Colleges of Technology, United Arabe of Emirates
Munir Naveed, Higher Colleges of Technology, United Arabe of Emirates
Suman Deb, NIT Agartala, India
Nirmalya Kar, National Institute of Technology Agartala, India
Ashim Saha, National Institute of Technology Agartala, India
Ibragim Suleimenov, Almaty University of Power Engineering and Telecommunications, Kazakhstan
Akhat Bakirov, Almaty University of Power Engineering and Telecommunications, Kazakhstan
Sebastian Heil, Chemnitz Technical University, Germany

Content from this work may be used under the terms of the Creative Commons Attribution 3.0 licence. Any further distribution of this work must maintain attribution to the author(s) and the title of the work, journal citation and DOI.
Published under licence by IOP Publishing Ltd

IOP Conf. Series: Materials Science and Engineering **946** (2020) 011003 doi:10.1088/1757-899X/946/1/011003

Virtual Disclaimer

4th International Conference on Information Processing and Control Engineering (ICIPCE 2020) was held in the virtual way at university of California, Berkeley, USA during June 27-29, 2020. Due to unexpected outbreak of Co-vid 19, considering the safety and the smooth publication procedures, the conference committee collected suggestions from attendees, resulting in the taking place of the online conference through Zoom.

With the e-conference, the speakers were talking 45 minutes each concerning their research field. The authors shared the 15-minute presentation with attendees.

The interactive conference ended successfully.

Content from this work may be used under the terms of the Creative Commons Attribution 3.0 licence. Any further distribution of this work must maintain attribution to the author(s) and the title of the work, journal citation and DOI.
Published under licence by IOP Publishing Ltd

IOP Conf. Series: Materials Science and Engineering **946** (2020) 011004 doi:10.1088/1757-899X/946/1/011004

Peer review declaration

All papers published in this volume of IOP Conference Series: Materials Science and Engineering have been peer reviewed through processes administered by the Editors. Reviews were conducted by expert referees to the professional and scientific standards expected of a proceedings journal published by IOP Publishing.

Type of peer review: Double-blind

Conference submission management system: ICONF.ORG

Number of submissions received: 27

Number of submissions sent for review: 18

Number of submissions accepted:9

Acceptance Rate (Number of Submissions Accepted / Number of Submissions Received X 100): 50%

Average number of reviews per paper: 2
Total number of reviewers involved: 10

Any additional info on review process:

Contact person for queries: Lucia Yang, icipce@saise.org

Content from this work may be used under the terms of the Creative Commons Attribution 3.0 licence. Any further distribution of this work must maintain attribution to the author(s) and the title of the work, journal citation and DOI.
Published under licence by IOP Publishing Ltd

IOP Conf. Series: Materials Science and Engineering **946** (2020) 012001 doi:10.1088/1757-899X/946/1/012001

Prerequisites for the analysis of the neural networks functioning in terms of projective geometry

Yelizaveta S Vitulyova[1], Akhat S Bakirov[1,2*], Dina B Shaltykova[3] and Ibragim E Suleimenov[2]

[1] Almaty University of Power Engineering and Telecommunications, Almaty, Republic of Kazakhstan
[2] National Academy of Engineering of the Republic of Kazakhstan, Almaty, Republic of Kazakhstan
[3] Institute of Information and Computational Technologies, Almaty, Republic of Kazakhstan

axatmr@mail.ru

Abstract. It is shown that during training an artificial neural network built on elements with a threshold activation function, the weighting coefficients of each neuron can be selected from a certain discrete set, the nature of which depends on the number of neurons inputs. An example of a neuron with three inputs is considered in detail. It is shown that there are only 8 non-trivial combinations of weight coefficients that specify the work of neurons of this variety: any other combination of weight coefficients that describe the work of a non-degenerate neuron with this number of inputs can be replaced by one of the above 8 combinations without changing the character of the neuron functioning in the network. A method has been developed that allows one to determine relevant combinations of weights for neurons with an arbitrary number of inputs. It is shown that the proposed approach creates the prerequisites for assessing the "information power" of neural networks based on data on the nature of the connections between the neurons that form it.

1. Introduction

There is an extensive class of tasks related to the classification of some objects according to one or another group of attributes, for the solution of which neural networks are used [1-5]. Such tasks de facto come down to finding an empirically determined rule, according to which a certain sequence of variables taking logical values assigned to an object from a set of images represented in digital form. We emphasize that this rule is found empirically in the process of training a neural network. Moreover, the existing training procedures for neural networks are guided by a continuous change in weighting coefficients. This, inter alia, is due to the fact that for the implementation of existing training procedures, as a rule, one has to use differentiable neuron activation functions.

At the same time, those neural networks that are used to solve the problems of image classification, in principle, could be solved using neural networks using threshold activation functions, which follows from the very nature of the problem (mapping an object described by a set of binary variables from one sets, to a similar object from another set).

In this paper, it is shown that analysis of neural networks` operation using the threshold activation function can be significantly simplified by using the ideas of projective geometry. It is shown that the set of weighting factors that specify the operation of a neuron with a threshold activation function can

Content from this work may be used under the terms of the Creative Commons Attribution 3.0 licence. Any further distribution of this work must maintain attribution to the author(s) and the title of the work, journal citation and DOI.
Published under licence by IOP Publishing Ltd

actually be selected from some discrete set, the nature of which is also established in this paper. This creates the prerequisites for a significant simplification of training procedures for artificial neural networks (ANN) that solve classification problems, as well as other tasks whose nature allows the use of ANNs whose elements have a threshold activation function.

There is also a well-defined problem, usually interpreted as "the neural networks logical opacity" [6]. In particular, in the vast majority of cases, the weights of a neural network are formed empirically in the process of its training using specific images. Algorithms that make it possible to judge why the weight coefficients formed during the training take exactly such and not other values, as a rule, remain unknown.

This does not create significant problems from the point of view of the applied use of ANNs, however, their logical opacity does not allow solving many problems that arise when trying to use the neural networks theory in other areas of knowledge.

So, the neural networks theory, in principle, can be used to interpret such concepts as "mentality", "Collective unconscious" [7] and other phenomena that describe society [8]. Indeed, let us consider two people entering into conversation with each other. It is customary to say that in this case two individuals communicate. However, this is nothing more than an approximation, and very rude. In fact, we are talking about the fact that the neurons that make up the brain of each interlocutor exchange signals with each other. From the point of view of the neural network functioning, it is absolutely not important what kind of nature those signals exchanged by neurons possess. Therefore, we can conclude that as a result of the dialogue, a certain common neural network is formed, consisting of two fragments, each of which is localized within the brain of two communicating individuals. Continuing this logic, it is easy to conclude that there is a global communication network that can be identified with the noosphere, understood in the spirit of V.I. Vernadsky`s point of view [9]. Moreover, there is every reason to believe that information that can only be stored indirectly in connection with the information recorded by the memory of individuals can be stored in an extensive neural network. It can be assumed that it is this information that determines the nature of what is called "collective unconscious", "dictate of the environment" in humanitarian literature, etc.

Today, statements of this kind are nothing more than philosophical conclusions arising from the logic of dialectical positivism [10-12]. Their verification by means of the theory of neural networks, however, creates quite certain prerequisites for the transition to a quantitative description of several processes taking place in human society. However, in order to proceed to quantitative conclusions, first of all, it is necessary to solve the problem of neural networks logical opacity.

The results of this work, in which it is shown that the theoretical description of neural networks with threshold activation functions can be significantly simplified, allows us to make a certain step in this direction.

2. The essence of the proposed approach

Let's start from the expression connecting the variable characterizing the output state of neuron Y with the variables characterizing the state of inputs X_i.

$$Y = \theta(w_1X_1 + w_2X_2 + \cdots + w_NX_N), \tag{1}$$

where w_i - weighting factors,

$$\theta(x) = \begin{cases} -1, & x \leq 0 \\ +1, & x > 0 \end{cases} \tag{2}$$

Relation (1) can be used to analyse the operation of neural networks of the type under consideration without loss of generality, since for a similar formula with a constant term a

$$Y = \theta(w_1X_1 + w_2X_2 + \cdots + w_NX_N + a) \tag{3}$$

one can formally increase the number of considered inputs per unit and put $w_{N+1} = a$, $X_{N+1} = 1$.

Without loss of generality, we can also assume that all variables X_i can only take discrete values of -1 and +1. (Often the case is considered when these variables acquire the values 0, + 1, but it reduces to the considered elementary change of variables.)

Formula (1) is easy to interpret from a geometric point of view. Indeed, it actually defines a section of a hypercube in N-dimensional Euclidean space by a hyperplane passing through the origin

$$w_1X_1 + w_2X_2 + \cdots + w_NX_N = 0 \quad (4)$$

Relation (1) indicates that the variable Y takes the value +1 if the product of the normal vector to a given hyperplane and the radius vector of one of the vertices of the hypercube is positive, and -1 in the opposite case. In other words, the hyperplane (4) divides the N-dimensional space into two regions, and the value of the variable Y depends on which one the corresponding vertex of the hypercube falls into.

Expression (4) can also be considered from the perspective of projective geometry [13], that uses the concept of dual space. Namely, record (4) is symmetric with respect to the sets of quantities $(w_1, w_2, \dots w_N)$ and $(X_1, X_2, \dots, X_N)$, i.e. it can be considered both as an equation of a hyperplane in the original space (then the set of quantities $(w_1, w_2, \dots w_N)$ is considered as specifying the orientation of this hyperplane), and as an equation of a hyperplane in a dual space. The orientation of this hyperplane in dual space is set, respectively, by a set of quantities $(X_1, X_2, \dots, X_N)$.

Accordingly, the set of quantities $(w_1, w_2, \dots w_N)$ has the meaning of the coordinates of a point in dual space. A point in the original space becomes a hyperplane in dual space, and a hyperplane becomes a point. For the case N = 3, this is illustrated by Fig. 1, which shows a cube in three-dimensional space, i.e. a set that exhausts the possible states of the inputs of a neuron described by function (1), and the figure into which it passes in dual space (regular octahedron). Six vertices of an octahedron correspond to six faces of a cube, and eight vertices of a cube correspond to eight faces of an octahedron.

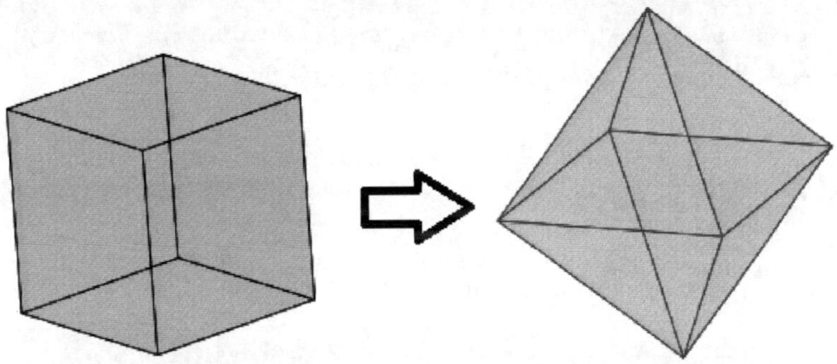

Figure 1. In dual space, a cube transforms into an octahedron.

Thus, a set of weighting factors that specify the operation of a neuron with a threshold activation function is represented by a point in a dual N-dimensional space. The set of vertices of an N-dimensional hypercube in a dual space becomes a set of hyperplanes, each of which also divides the dual space into two parts.

We say that this set of weights $\{w_i\}$ belongs to the set $\Gamma^+\left(\{X_i\}|_{i=1}^{i=N}\right)$,

$$(w_1, w_2, w_3, \dots, w_N) \in \Gamma^+\left(\{X_i\}|_{i=1}^{i=N}\right) \quad (5)$$

if the condition is met

$$\sum_i^N w_iX_i > 0 \quad (6)$$

and, conversely, it belongs to the set $\Gamma^-\left(\{X_i\}|_{i=1}^{i=N}\right)$ if the condition is met

$$\sum_i^N w_iX_i \le 0 \quad (7)$$

For a hypercube of dimension N there exist exactly 2^{N-1} sets $\Gamma^+\left(\{X_i\}|_{i=1}^{i=N}\right)$ that do not coincide with any of the sets $\Gamma^-\left(\{X_i\}|_{i=1}^{i=N}\right)$, and exactly 2^{N-1} of such sets $\Gamma^-\left(\{X_i\}|_{i=1}^{i=N}\right)$. This is due to the fact that for each vertex defined by the set of binary variables $X_i \in (-1,1)$, i.e. vector

IOP Conf. Series: Materials Science and Engineering **946** (2020) 012001 doi:10.1088/1757-899X/946/1/012001

$$\vec{n} = (X_1, X_2, X_3, \dots, X_N) \tag{8}$$

It is possible specify a vertex that is characterized by an oppositely directed vector

$$\vec{n} = -(X_1, X_2, X_3, \dots, X_N) \tag{9}$$

Let's form sets

$$\Lambda(\vec{\beta}) = \cup_{\vec{\beta}} \Gamma^{\beta_j}(\{X_i\}|_{i=1}^{i=N}) \tag{10}$$

where the multi-index $\vec{\beta}$ denotes the sequence of the choice of signs of the sets $\Gamma^{\pm}(\{X_i\}|_{i=1}^{i=N})$ of length 2^{N-1}.

Let the functioning of a neuron with N inputs be described by the activation function (2). Then, the result of applying to the inputs of any sequences N binary variables for two different activation functions described by row vectors $\vec{w}^{(1)} = (w_1^1, w_2^1, w_3^1, \dots, w_N^1)$ and $\vec{w}^{(2)} = (w_1^1, w_2^1, w_3^1, \dots, w_N^1)$ will be the same if $\vec{w}^{(1)} \in \Lambda(\vec{\beta}), \vec{w}^{(2)} \in \Lambda(\vec{\beta})$, i.e. these vectors belong to the same set $\Lambda(\vec{\beta})$. In other words, neurons connected to the network with weights $\vec{w}^{(1)}$ and $\vec{w}^{(2)}$ work indistinguishably.

This follows from the fact that for two vectors from this set their scalar products by vectors (10) corresponding to any possible combination of binary variables supplied to the inputs of a neural network will have the same sign.

This proves the following theorem.

Theorem 1

In the formula describing the activation of a neuron (2) when applying to its input a set of binary signals capable of taking values -1 and +1, the vector of weighting coefficients $(w_1, w_2, w_3, \dots, w_N)$ *belonging to the set* $\Lambda(\vec{\beta})$ *can be replaced by any other vector from this set.*

Formally, such sets can be formed $2^{2^{N-1}}$, i.e. there are no more than $2^{2^{N-1}}$ different vectors describing possible variants of sets of weight coefficients that are de facto distinguishable. In reality, the very nature of function (2) leads to the fact that there are much fewer possible options.

We show this by the example of a neuron with three inputs.

For the case of a neuron with three inputs, the activation function has the following form.

$$Y = \theta(w_1 X_1 + w_2 X_2 + w_3 X_3) \tag{11}$$

Based on this function, it can be argued that all possible areas in the projective space, which will differ in the nature of the reaction of the neuron to the state of the input variables, are determined by the position of the following four planes.

$$w_1 + w_2 + w_3 = 0 \tag{12}$$

$$w_1 + w_2 - w_3 = 0 \tag{13}$$

$$w_1 - w_2 + w_3 = 0 \tag{14}$$

$$w_1 - w_2 - w_3 = 0 \tag{15}$$

In other words, these four planes, formulas (12)-(15) divide the entire dual space into well-defined subdomains. If two sets of weights belong to the same area, then, as shown above, the corresponding neurons will behave identically.

In other words, as follows from what was said above, there are as many possible variants of the implementation of neurons as there are subdomains into which the dual space is divided by planes defined by equations (12)-(15).

For the three-dimensional case under consideration formally, there should be 16 such regions. In reality, there are exactly fourteen such subdomains; this is due to the fact that some intersections of the planes defined by equations (12) - (15) are empty sets.

It is possible to choose the following values of vectors characterizing the subdomains into which the dual space splits.

$$\vec{w_1} = \pm(1,1,1), \vec{w_2} = \pm(1,1,-1), \vec{w_3} = \pm(1,-1,1), \vec{w_4} = \pm(1,-1,-1) \quad (16)$$

$$\vec{w_5} = \pm(1,0,0), \vec{w_6} = \pm(0,1,0), \vec{w_7} = \pm(0,0,1), \vec{w_8} = \pm(0,0,0) \quad (17)$$

In the list of such vectors in formulas (16) - (17), a vector $\vec{w_8}$ is also formally added, all of whose components are zero. This subdomain is a degenerate subdomain that is formed when the intersection of some sets $\Gamma^{\pm}(\{X_i\}|_{i=1}^{i=N})$ defined above is empty. In other words, there are no sixteen transfigurations, as follows from the formal calculation given above, but fourteen. Two regions are degenerate and correspond to the fact that the neuron is actually absent, i.e. all components of the vector are equal to zero.

The nature of the vectors set defined by formulas (16)-(17), which can be called basic, is illustrated by Figure 2.

Figure 2. Correspondence of vectors from the set (16), (17) to the cuboctahedron faces

Further, as follows directly from formulas (17), eight out of 16 variants are actually degenerate. Indeed, if only one weighting factor in the considered formula is nonzero, then the neuron de facto does not have three inputs, but only one. The cases corresponding to formulas (17) also correspond to situations when two of the three inputs are not connected to the circuit.

Thus, we have proved that any non-degenerate neuron that has three inputs and a stepwise activation function can in fact be described through weighting factors, which also take only -1 and +1 values. In other words,

This result admits a well-defined generalization, more precisely, it holds

Theorem 2

For any non-degenerate neuron with an odd number of inputs, the set of weighting coefficients can be replaced by one of the sequences of the form $(q_1, q_2, \dots, q_N)$, the elements of which can only take values -1 or +1.

By "non-degenerate" in this formulation we mean a neuron, none of which weights are equal to zero. The proof of the theorem is given in the same way as was used for the case N = 3.

This theorem suggests that the weights of non-degenerate neurons with an odd number of inputs and a threshold activation function are de facto described by vectors from the same set as the sets of input variables. This set can also be associated with the vertices of a hypercube with edge 2 in an N-dimensional space.

The case of even N is somewhat more complicated for analysis, but even in this case it is possible to show that the real number of possible sequences of weighting coefficients is relatively small.

3. Conclusion

Thus, the results of this work show that instead of arbitrary combinations of weighting coefficients describing the operation of neural networks with threshold activation function, we can take combinations of weighting coefficients whose elements acquire discrete values taken from a well-defined discrete set.

Outside the scope of the work, there remains the question of training algorithms for neural networks with threshold activation functions and sets of weighting coefficients selected from certain fixed sets. It is solved using the method of "logical differentiation", the consideration of which is beyond the scope of this work.

The most significant conclusion of this work, however, is that the obtained results allow us to estimate the amount of information that is stored in the weight coefficients of the neural network. The possibility of going over to discrete sets clearly indicates that Shannon's formulas are applicable to its calculation. Thus, prerequisites are created in order to try, for example, compare the amount of information stored in a collective neural network (noosphere) with the amount of information directly related to the memory of individuals. Such a comparison, obviously, represents the first step towards building quantitative models that describe human society from the standpoint of the neural networks' theory. The relevance of creating such models is not in doubt due to the increasing role of telecommunications in society and the actual formation of complex human-machine systems.

4. References

[1] Basheer I A, Hajmeer M 2000 Artificial neural networks: fundamentals, computing, design, and application *Journal of microbiological methods* **43 (1)** pp 3-31 DOI: https://doi.org/10.1016/S0167-7012(00)00201-3

[2] Ibragim S. et al. 2020 Example of the use of artificial neural network in the educational process *Future of Information and Communication Conference, Springer, Cham* pp 420-430 DOI: https://doi.org/10.1007/978-3-030-39445-5_31

[3] Romanovsky A V and Shokin Ya V 2010 Neuroeconomics and its integration into economic science. *Economics* **70 (9)** pp 42-44

[4] Ciregan D, Meier U, Schmidhuber J 2012 Multi-column deep neural networks for image classification 2012 *IEEE conference on computer vision and pattern recognition* pp 3642-49 DOI: 10.1109/CVPR.2012.6248110

[5] Liang M, Hu X 2015 Recurrent convolutional neural network for object recognition *Proceedings of the IEEE conference on computer vision and pattern recognition* pp 3367-75

[6] Burrell J 2016 How the machine 'thinks': Understanding opacity in machine learning algorithms Big Data & Society 3(1) DOI: https://doi.org/10.1177/2053951715622512

[7] Gunderson K 1985 Mentality and machines *University of Minnesota Press*

[8] Suleimenov I and Bakirov A 2018 Semi-empirical Theory of Distributions of Users of Social Online Networks by the Number of Contacts *MATEC Web Conf.* **214** DOI: 10.1051/matecconf/201821402001

[9] Vernadsky V I 2013 A few words about the noosphere *Noosphere studies* **1** pp 6-17

[10] Suleimenov I, Massalimova A, Bakirov A and Gabrielyan O 2018 Neural Networks and the Philosophy of Dialectical Positivism *MATEC Web Conf.* **214** DOI: 10.1051/matecconf/201821402002

[11] Suleimenov I E, Gabrielyan O A, Bakirov A S and Vitulyova Ye S 2019 Dialectical Understanding of Information in the Context of the Artificial Intelligence Problems *IOP Conference Series: Materials Science and Engineering* **630** DOI: 10.1088/1757-899X/630/1/012007

[12] Gabrielyan O, Suleimenov I, Vituleva E 2019 Artificial intelligence in the context of noosphere studies *SCTCMG 2019 International Scientific Conference «Social and Cultural Transformations in the Context of Modern Globalism»* https://doi.org/10.15405/epsbs.2019.12.04.130

[13] Coxeter H S M 2003 Projective geometry *Springer Science & Business Media*

IOP Conf. Series: Materials Science and Engineering **946** (2020) 012002 doi:10.1088/1757-899X/946/1/012002

Application of ternary logic for digital signal processing

I Moldakhan[1], D B Shaltikova[2], Z M Egemberdyeva[1] and I E Suleimenov[2]

[1] Almaty University of Power Engineering and Telecommunications, 126/1 Baitursynova Street, Almaty, Kazakhstan
[2] Institute of information and computational technologies, 125 Pushkina Street, Almaty, Kazakhstan

imoldakhan@gmail.com

Abstract. The article shown that digital processing of slowly changing signals based on ternary logic has significant advantages compared to processing based on binary logic. It is shown that for slowly changing signals, digital processing can be performed on the basis of dividing analysed signal into frequency bands. In particular, information on derivatives of a slowly varying signal which spectrum lies in the frequency band from 0 to ω_0, can be restored basing on the analysis of its component lying in the frequency band from $\omega_0/3$, to ω_0..

1. Introduction

There are a number of reports [1-3] which prove that in relation to digital signal processing ternary logic has a number of advantages compared to binary. The main advantage is a reduction of operations' numbers required to convert an analogue signal into digital form. According to available estimates, the number of operations required to convert a signal into digital form using ternary logic is approximately in 1.5 times less than operations numbers that must be performed to carry out a similar procedure using binary logic.

However, the most clearly examples showing benefits of ternary logic over binary are related to digital processing of signals that relatively slowly change in time (more precisely with processing of signals which time derivative does not exceed a certain value).

For such signals can be formed ε-cover. In other words it is possible to choose Δu_o subinterval by levels when a signal value $\Delta u_{(i+1)}$ on the $(i+1)$-st beat will differ from value of Δu_i on the i beat no more than Δu_o. That is if the signal on $(i+1)$-st beat corresponded to the discrete level number j then the signal on the $(i+1)$-st beat will correspond the level with one of numbers $j-1, j, j+1$.

Obviously, an amount of information contained in such signal is most conveniently measured in trit (unit of information [4], which name is built by analogy with the name of generally accepted unit "bit", but relating not to binary, but to ternary logic). Obviously, that information quantity which exhaustively describes signals of the type under consideration is an equal to N trit, where N - is a number of beats (with the accuracy to information which contained in an initial value of signal).

However, by historical reasons, ternary logic has not become widespread yet, although there are a number of tasks that require digital processing of slowly changing signals. An example on this regard is any system which applied to track values that are changing relatively slowly in time (in particular, it refers to any systems which built on measurements of inertial quantities, f.e. the temperature of sufficiently massive bodies [5]). Also, a typical example are adaptive optics systems which intend for use in the solar energy industry. Specifically, to track a position of the Sun in the sky [6-8]. In this article provides an additional evidence of the benefits of ternary logic. More precisely, it is shown that the

Content from this work may be used under the terms of the Creative Commons Attribution 3.0 licence. Any further distribution of this work must maintain attribution to the author(s) and the title of the work, journal citation and DOI.
Published under licence by IOP Publishing Ltd

digital processing of signals which have ε-coverage, in terms of ternary logic, can be performed using frequency filters (that is, using the separation of the original signal in frequency ranges).

2. Problem statement

Currently, digital methods are widely used for signal processing. In the vast majority of cases, digital signal processing uses binary logic, which is why analog-to-digital converters are used. These converters must be applied in the same way both to the fast-changing signal and to the slow-changing signal, which was discussed above. Using such converters does not allow you to use the fact that the derivative of these signals is limited. The processing of each clock cycle proceeds independently. This leads to the fact that the existing types of analog-to-digital converters require a significant number of operations, as follows from the materials of this work. If we take into account the specifics of slowly changing signals, the number of operations that are necessary to translate them into the digital form will be significantly reduced [9-13].There is every reason to believe that reducing the number of operations in digital signal processing is becoming an increasingly urgent task due to the development of the Internet of things [14], big data [15] and digitalization of the economy as a whole [16].
The question of further development of ternary logic is also relevant from a philosophical point of view, in particular, for dialectical positivism [17,18]; within the framework of modern mathematical logic, very non-trivial logics are also considered [19,20], the construction of which differs significantly from the logic of Aristotle and Buhl.

3. Prerequisites for development of a spectral method for digital processing of slowly varying signals

Obviously, that if the signal is approximately replaced by its model, consistent to ε-cover, then it automatically supposed using of discrete beats (time intervals) during which the value of signal is considered constant. Duration of all beats without losing of generality can be considered as the same.
A spectrum of such model signal is intentionally limited. Specifically, it belongs to a range from 0 to $\frac{1}{2T_0}$ by a frequency scale, where T_0 - is a duration of separate beats.
A signal $u(t)$ with a limited spectrum can be presented as follows

$$u(t) = \frac{1}{2\pi}\int_{-\omega_0}^{\omega_0} exp(i\omega t)A(\omega)d\omega \tag{1}$$

where ω - angular frequency, $A(\omega)$ - signal spectrum, this function takes complex values, ω_0 - liminal angular frequency.
At the same time, if the signal changes slowly then to describe it in a vicinity of some selected point (of point in time) correctly representation through a Taylor series is fair to use

$$u(t_i \pm \Delta t) = u(t_i) \pm \left.\frac{du}{dt}\right|_{t=t_i} \Delta t + \left.\frac{d^2u}{dt^2}\right|_{t=t_i} (\Delta t)^2 \tag{2}$$

which within transition to discrete quantity take the form

$$u_{i\pm1} = u_i \pm \left.\frac{du}{dt}\right|_i \Delta t_0 + \left.\frac{d^2u}{dt^2}\right|_i (\Delta t_0)^2 \tag{3}$$

Considering that the values of $u\_(i \pm 1)$ for the signals of concerned type can differ from u_i no more than to one discrete level and the value of Δt_0 - is a constant, so using of formula (3) de facto means that it should be found some procedure of assessment of the first and second derivative. In essence, representation of the following form is constructed

$$\left(\pm\left.\frac{du}{dt}\right|_i, \left.\frac{d^2u}{dt^2}\right|_i\right) \to q \tag{4}$$

where q can take values -1, 0, +1.
From the formula(4), even without detailed calculations, follows that while considering a discrete signals in expansion into Taylor's series - it is necessary to consider elements up to the second derivative inclusive. Because using only the first derivative will not reflect, for example, a situation where both the value $u\_(i+1)$ and the value $u\_(i-1)$ are excess of one level the value of the signal u_i.
The question arises - exist there a connection between conceptions one of the same discretely varying signal through formula (1) and through formula (2). In other words, is it possible to determine characteristics of displaying (4) by analyzing frequency spectrum of the signal?

IOP Conf. Series: Materials Science and Engineering **946** (2020) 012002 doi:10.1088/1757-899X/946/1/012002

4. Evidence of applicability of a spectral method of digital processing to slowly varying signals

Let us divide a spectral interval, in which the spectrum of the function under consideration is defined, into the following three subintervals $\left[-\omega_0, -\frac{1}{3}\omega_0\right]$, $\left[-\frac{1}{3}\omega_0, \frac{1}{3}\omega_0\right]$ and $\left[\frac{1}{3}\omega_0, \omega_0\right]$.

Let us calculate contribution to the integral (1) for each of these intervals individually. For interval $\left[\frac{1}{3}\omega_0, \omega_0\right]$, changing a variable

$$\omega_1 = \omega - \frac{2}{3}\omega_0 \tag{5}$$

we have

$$u_1(t) = exp\left(i\frac{2}{3}\omega_0 t\right)\int_{-\frac{1}{3}\omega_0}^{\frac{1}{3}\omega_0} exp(i\omega_1 t)A(\omega_1)d\omega_1 = exp\left(-i\frac{2}{3}\omega_0 t\right)\tilde{u}_1 \tag{6}$$

Obviously, the integral in formula (6) in its form looks identical to the integral (1) i.e. it also describes a signal, which is equivalent to a system of equidistant counting, but following with period in three times upper $T_1 = 3T_0$.

Considering an interval $\left[-\omega_0, -\frac{1}{3}\omega_0\right]$ in the same way let us to make a change of variable

$$\omega_3 = \omega + \frac{2}{3}\omega_0 \tag{7}$$

one can obtain

$$u_2(t) = exp\left(-i\frac{2}{3}\omega_0 t\right)\int_{-\omega_0}^{-\frac{1}{3}\omega_0} exp(i\omega_3 t)A(\omega_3)d\omega_3 = exp\left(-i\frac{2}{3}\omega_0 t\right)\tilde{u}_2 \tag{8}$$

where the period of sequence of counts of signal defined by the integral appearing on the right side of (8) is also tripled comparing to the original $T_1 = 3T_0$

For interval $\left[-\frac{1}{3}\omega_0, \frac{1}{3}\omega_0\right]$ the formula (1) is applied directly and here also $T_1 = 3T_0$ takes place.

$$u_3(t) = \frac{1}{2\pi}\int_{-\frac{\omega_0}{3}}^{\frac{\omega_0}{3}} exp(i\omega t)A(\omega)d\omega \tag{9}$$

Summing up formulas (6), (8) and (9), we obtain the following representation for the initial signal

$$u(t) = \tilde{u}_1(nT_1)exp\left(-i\frac{2}{3}\omega_0 t\right) + \tilde{u}_3(nT_1)exp\left(-i\frac{2}{3}\omega_0 t\right) + u_2(nT_1) \tag{10}$$

where all functions $\tilde{u}_1$, $\tilde{u}_3$, u_2 have a limited spectrum in a frequency band up to $\frac{1}{3}\omega_0$. This, in particular, means that for three beats of the initial signal, there is only one count of these functions. In other words each of them can be associated with a value that will remain unchanged for three beats.

Comparing the result obtained in (10) with the formula (3) which directly derives from expansion into Taylor series, it is possible to see that information on the value of the second and first derivatives actually turns out to be in the frequency band from $\frac{1}{3}\omega_0$ to ω_0, which can be separated from a frequency band that allows to restore the function u_2 (and this can be done using typical electronic tools).

We emphasize that considerations forcing to take into account precisely the first two derivatives are fundamental. That is why the range of frequencies in which the spectrum of the signal considered above fits was divided into two unequal parts.

Let us consider a harmonic signal modulated by a signal in a certain frequency band. As is well known from radio engineering, in this case a signal spectrum contains combination frequencies, in other words, a band occupied by this signal becomes twice as wide as the band of the original signal. Nevertheless, information transmitted by such signal corresponds to a width of the original band, since in this case opportunities of the double bandwidth capability is not fully utilized. Along with modulation of the harmonic signal, let us say sinusoidal, in order to transmit information it is possible to use modulation of harmonica signal opposite in phase, i.e. in this case cosine one. This signal is capable of carrying the same information, in other words, information can be transmitted in a doubled frequency band by modulating two antiphase signals.

For that very reason, we used a quite specific division of frequency band into two unequal parts, related to each other in width as two to one. It can be seen that in this case information on signal is actually contained in three frequency bands, each of which is 1/3 of the original. However, since bands from

$\frac{1}{3}\omega_0$ to $\frac{2}{3}\omega_0$ and from $\frac{2}{3}\omega_0$ to ω_0 are actually "mixed", it is needed to combine and consider them together, concurrently taking into account the first and second derivatives.

The diagram shown in Fig. 1 illustrates how information obtained by radio-frequency analysis of the first and second derivatives can be used to restore behavior of a function with ε-cover over a period of three beats.

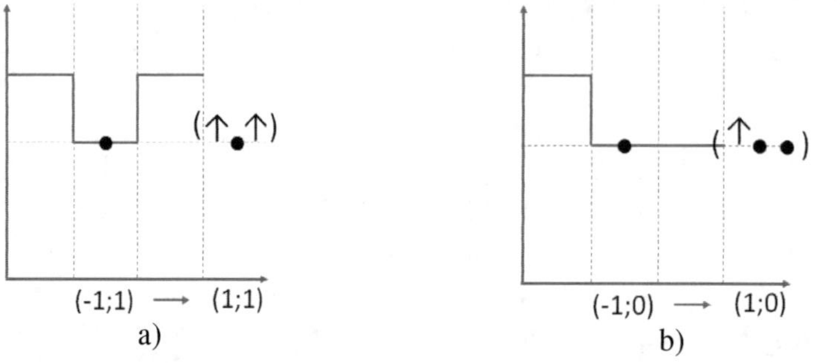

Figure 1. Illustration of decoding of relationship between a nature of coding sequences for a three-stroke range and signs of the first and second derivative; examples of coding sequences (-1, 1) (a) and (-1,0) (b)

This change can obviously be encoded in a sequence of two symbols, each of them can take the value 1, 0 and - 1.

These symbols correspond to changes in the signal level during transition from the first beat to the second and from the second one to the third beat (three possible particular cases are shown in Fig. 1). The same figure also shows examples of circuits illustrating the nature of deviation of the signal from the center beat (the right parts of figures). For example, the first signal with a change of the signal in accordance with the -1, 1 sequence, is schematically represented by drawing (the right part of Fig.1a), on which there are two arrows pointing up. This means that deviation of the signal from the value that it takes on the central beat is positive for the beat located on the left and for the beat located on the right. Nine possible combinations of a sequence of change of levels correspond to nine possible combinations of values of the second and first derivatives, which are assigned the value either +1, or 0, or -1.

Exactly this consideration can be taken as a basis for digital radio engineering processing of slowly changing signals (more precisely, signals with ε-cover). Namely, as follows from considerations above, in order to establish the nature of the signal change over three selected beats it is sufficient to identify only the signs of the first and second derivatives.

Thereby, dividing the signals into frequency bands allows to highlight the behavior of the signal over three beats by analyzing the sign of the signal in a well-defined frequency range. Moreover, to highlight the sign of cosine and sinusoidal signals, it is quite possible to use standard radio engineering techniques.

5. Conclusion

Hereby, digital processing of signals based on ternary logic has really very serious advantages compared to binary, especially if we consider signals with ε-cover. In this case, comprehensive information about derivatives of a given signal in a group of three beats can be established basing on an analysis of its component lying in a frequency band from $\omega 0/3$ to $\omega 0$, where $\omega 0$ - the frequency corresponding by the time to discretization step of the original signal. Wherein, in order to restore the signal, it is sufficient to establish values of the derivatives in digital form i.e. correlate them with values corresponding to a ternary logic: "-1", "0", "+1".

The proposed approach can be implemented in hardware since its implementation is essentially based on conventional spectral methods. It is enough to select the signals in the corresponding frequency band and determine whether they exceed the specified threshold or not. Thus, further work in this direction will significantly simplify the digital processing of slowly changing signals. We emphasize that we do not intend to contrast ternary logic with binary logic in this work. On the contrary, we can predict that in the future there should be flexible platforms that, depending on the need, will use one or another logic

IOP Conf. Series: Materials Science and Engineering **946** (2020) 012002 doi:10.1088/1757-899X/946/1/012002

to reduce the number of operations, which is an urgent task in the modern world due to the development of the digital economy.

6. References

[1] B Ay, S C Crist, Design and evaluation of a bipolar tristate logic family, IEEE J. Solid-State Circuits, vol. SC-17, pp. 16-19, Feb. 1982.

[2] B Hayes, Third base, American Scientist, 2001, v. 89, ntist, , e, eb..

[3] A S Korotkov, D V Morozov, M M Pilipko, A Sinha, Delta-Sigma ADC for Ternary Code System (Part I: Modulator Realization), Proc. Int. Symp. Signals, Circuits and Systems, July 2007, vol. 1, p. 1-4.

[4] D M Klionsky, N I Oreshko, V V Geppener, Empirical mode decomposition in segmentation and clustering of slowly and fast changing non-stationary signals, Representation, Processing, Analysis, and Understanding of Images, First Online, March 2009

[5] I V Pershin, Calculation of thermal fields of massive bodies heated by a radial heat source in the three-dimensional cas, Trudy Inst. Mat. i Mekh. UrO RAN, 23, AN,N,RAN,,, i Mek

[6] Jifeng Song., Yongping Yang., Yong Zhu., and Zhou Jin, A high precision tracking system based on a hybrid strategy designed for concentrated sunlight transmission via fibers, Renewable energy, 2013, p. 12-19

[7] Y Rizal et al. Application of solar position algorithm for sun-tracking system, Energy Procedia, vol. 32, pp. 160-165. 2013.

[8] P Roth , A Georgiev and H Boudinov. Design and construction of a system for sun-tracking, Renewable energy, vol.29(3), p. 393-402, 2004.

[9] M Mukaidono and M Regular, Ternary logic functions suitable for treating ambiguity, IEEE transactions on computers, vol.2, p. 179-183, 1986.

[10] Y Yamamoto and M Mukaidono, Meaningful special classes of ternary logic functions-regular ternary logic functions and ternary majority functions, IEEE transactions on computers, vol. 37(7), p. 799-806, 1988.

[11] T Felicijan and S Furber, An asynchronous ternary logic signaling system, IEEE Transactions on Very Large Scale Integration (VLSI) Systems, vol. 11(6), p. 1114-1119, 2003.

[12] R Mariani et al., A useful application of CMOS ternary logic to the realisation of asynchronous circuits, Proceedings 1997 27th International Symposium on Multiple-Valued Logic. IEEE, p. 203-208, 1997.

[13] C Kooi and J Weaver, Non-linear Hall-effect ternary logic element, Solid-State Electronics, vol. 7(5), p. 311-321, 1964.

[14] L Tan and N Wang, Future internet: The internet of things,. In 2010 3rd international conference on advanced computer theory and engineering (ICACTE), vol. 5, pp. 376-380), 2010.

[15] A McAfee, E Brynjolfsson, T Davenport, D Patil and D Barton, Big data: the management revolution, Harvard business review, vol. 90, p. 60-68, 2012.

[16] M Afonasova, M Panfilova, E Galichkina and B Ślusarczyk, Digitalization in economy and innovation: the effect on social and economic processes. Polish Journal of Management Studies, vol. 19, 2019

[17] I Suleimenov, A Massalimova, A Bakirov and O Gabrielyan, Neural Networks and the Philosophy of Dialectical Positivism. MATEC Web Conf. 214, 2018. DOI: 10.1051/matecconf/201821402002

[18] I Suleimenov, O Gabrielyan, A Bakirov and Ye Vitulyova, Dialectical Under-standing of Information in the Context of the Artificial Intelligence Problems. IOP Confer-ence Series: Materials Science and Engineering 630, 2019. DOI: 10.1088/1757-899X/630/1/012007

[19] A Voronkov, Theorem proving in non-standard logics based on the inverse method, International Conference on Automated Deduction, Springer, Berlin, Heidelberg, p. 648-662, 1992.

[20] J Rosser and H Wang, Non-standard models for formal logics, The journal of symbolic logic, vol. 15, pp. 113-129, 1950.

IOP Conf. Series: Materials Science and Engineering **946** (2020) 012003 doi:10.1088/1757-899X/946/1/012003

HGM: A Novel Monte-Carlo Simulations based Model for Malware Detection

M Naveed, M Alrammal, A Bensefia

Higher Colleges of Technology, Computer Information System Department, Abu Dhabi, UAE

{malrammal, mnaveed, abensafia}@hct.ac.ae

Abstract. Malware detection is a challenging and non-trivial task due to ever increase in several attacks and their sophistication level. Detection of such attacks demands the exploration of new approaches to generalize the attack patterns. One such approach is the use of Monte-Carlo simulations to train a reinforcement learning model. In this paper, we propose a self-adaptive Monte-Carlo simulation-based reinforcement model called Heuristic-based Generative Model (HGM), which generalizes the attack patterns in such a way that the new unknown attacks can be detected and flagged in real-time. The results show that HGM can detect a variety of malware with high accuracy.

1. Introduction

The Internet plays a primary rule in our life. Internet services like e-banking and e-commerce are used by a remarkable number of individuals and organizations. Such services encourage cyber-criminals to achieve their malicious intention by using Malware (malicious programs) [1]. Hundreds of millions of new samples of malicious software have been created. This indicates that approximately one million new threats/attacks are released daily [2].

Several anti-malware solutions have been proposed by security companies to protect organizations and individuals. These solutions are based on a repository where all the "known" types/definitions of malware are referenced; thus, they can be identified and blocked. However, due to the enormous increase in new malware samples, referencing all the malware has become a complicated task. This situation has created a need for intelligent approaches, flexible and adaptable, so that they can identify and block even "unknown" malware.

Machine learning approaches have shown real improvement and efficiency in analyzing and detecting malware. Most approaches are based on the selection of the best of features that describe best a malware, coupled to an appropriate machine learning algorithm [3-13]. In this paper, we extend our previous approach RCBM, based on reinforcement learning [14], to propose a new one called Heuristic-based Generative Model (HGM) that combines the best features of MOCART [15] and Random Forest [4-6].

The remaining of this paper is structured as follows: Section 2 discusses the different approaches used to detect and analyse malware. Section 3 explains the improvements made on our previous approach RBCM [14] to avoid converging to local-minima in the search spaces with a small range of values in an observation dataset. Sections 4 illustrates the experimentation setup and compares the performance of HGM with the state-of-work approaches. In sections 5, we present our conclusion and future work.

Content from this work may be used under the terms of the Creative Commons Attribution 3.0 licence. Any further distribution of this work must maintain attribution to the author(s) and the title of the work, journal citation and DOI.
Published under licence by IOP Publishing Ltd

IOP Conf. Series: Materials Science and Engineering **946** (2020) 012003 doi:10.1088/1757-899X/946/1/012003

2. Related Work

Machines learning approaches. proposed in the literature, can be organized in three major dimensions, according to Ucci's taxonomy [16]: malware's objective, malware's features, and the machine learning algorithm used. Since our study focuses on the Machine Learning relevance in the malware detection, we will focus on the third dimension mainly. Even, if the existing approaches have explored plenty of machine learning techniques, the most used ones are: Random Forest [4-6], Decision trees [7-8], Neuronal Network [9-11], Naïve Bayes [12-13], k-NN and SVM[13]; an exhaustive list is discussed in the following.

Several malware classification approaches have used the Random Forest clustering technique [4-6], due to it competitive performances compared to other algorithms. Karanja et al. [17] conducted a comparison between three clustering techniques, where he showed that the Random Forest overpasses the Naïve Bayes and the KNN algorithms; to this end he modelled the malwares as grayscale images. In another notable work presented in [6], the authors developed a malware clustering system: AMICO, evaluated thanks to a Random Forest technique, during nine months. This system was built around three main components: a download reconstruction module, which aims to reconstruct a real-time TCP traffic flow, where portable executable (PE) files were identified, and their reconstructed response copied by the download history module. Finally, a provenance classifier module analyses the behaviour of how the suspicious file was downloaded using the following set of features: the past file download, domain features, Server IP features, URL features, and the download request features. The training set of labelled malwares was monitored during two periods of time (one month and two months), and the experiments were conducted on the academic network for nine months and showed that AMICO was able to identify malware in 90% of the cases correctly.

Decision trees have also been considered in targeting malwares [7]. A Malware Targeting Recognition system (MaTR) was developed by combining decision tree algorithm and static heuristic features. In this study, the authors used over 100 heuristic features based on structural information (file attributes: name, path, attributes, size,..) and structural anomalies (section names, entry point, imports, exports,..). The system was trained with VX-Heaven dataset, and the decision tree classifier identified 99% of unknown malwares correctly, as announced by the authors.

Neuronal Network haven't been considered widely in malware detection approaches as the other techniques stated above [9-11]. In Andrade et al. [9], the authors proposed a new public dataset, that was evaluated using a recurrent neuronal network (RNN) classifier. This RNN differs from other RNN by the formation of direct cycles between their neurons. The demand of greater memory of this classifier was resolved by the usage of Long Short-Term Memory (LTSM). The approach performed a good identification rate in 67% of the cases.

The state-of-the-art approaches demonstrated that Random Forests approach is promising. It outperforms other approaches like the Naïve Bayes and KNN. It is an ensemble model of decision tree which extracts efficiently the correlation-based rules. As for MOCART, it is a real-time Monte-Carlo simulation model that trains reinforcement learning structure using a generative model. The approach has been successfully applied in partially observable domains with high uncertainty in search space. The main goal of this paper is to extend our previous approach RBCM [14] in such a way that it will combine the best features among Random Forest and MOCART in the generative model of RBCM. The new approach will be compared with the state-of-the-art approaches to test its accuracy and performance.

3. Heuristic-based Generative Model

HGM is a Monte-Carlo simulation-based model which uses the best features of RBCM and Random Forest. The generative model uses a novel sampling approach which is guided by a heuristic to search for a quality solution under real-time constraints. HGM also self-tunes the parameters for heuristic search.

HGM addresses the random unguided sampling-based search in RBCM which can find solutions with low accuracy in domains with low branching factor as shown in [14]. The high-level description of HGM generative model is shown below (Procedure HGM) for a domain D, sample space S and simulation length t, extension n. E is the error function.

IOP Conf. Series: Materials Science and Engineering **946** (2020) 012003 doi:10.1088/1757-899X/946/1/012003

Procedure HGM(D, t, n, S, E):

1. Generate n samples at s
2. Evaluate Learning Structure on n samples
3. Measure $E(s)$ for n samples
4. If $E(s, d) < \theta$
5. Generate $n+t$ samples
6. Update Learning Structure
7. Evaluate Learning Structure on $n+t$ sample
8. $\theta = E(s, d)$

The generative model of HGM extends the simulation length by $d+n$ as shown above in procedure HGM. The update decision is made by using heuristics $E(s)$ boundary *i.e.* θ. which depends on the current value of Q function which is a learning structure used in [14]. The θ value also represents validation of Q function on unseen data. HGM simulation model keeps the decision parameter dynamic and self-tuning to balance the trade-off between exploration and exploitation. Figure 1 explains the sampling strategy to guide the search along the path to quality solutions.

The search space at node *n1* is simulated with three transitions at three nodes *n2*, *n3* and *n4*. HGM simulates each sample strategy and evaluates the simulated paths so produce $E(s)$ for each solution quality. *n2* brings in minimum $E(s)$ than of its siblings and provides a better heuristic for selection of direction to explore further. This also helps the learning structure to tune its weight values that reduce the error of the learner. The heuristic value is self-tuned on each extension of search depth, for example, $E(s)$ at *n3* will be 0.002. If no siblings of *n1* produces lower error values than n1 the look-ahead search will terminate at this state. This sample strategy is intuitive to take the learner out of local-minima as $E(s)$ will never be smaller than the best of local-minima and generative model will explore more solutions.

Due to self-tuning of the heuristic function $E(s)$ in generative model, the model can converge to biased strategies which always select exploration of the solution space instead of exploiting the best solution found yet. However, to counter this issue, the extensions to look ahead search remains limited to a fixed number. This will reduce the probability of jumping out of a promising region in the solution space. The adaptive use of $E(s)$ also adds to the benefit of avoiding visiting the same sample space for more than once. This feature optimizes the time of searching the best solution in space S.

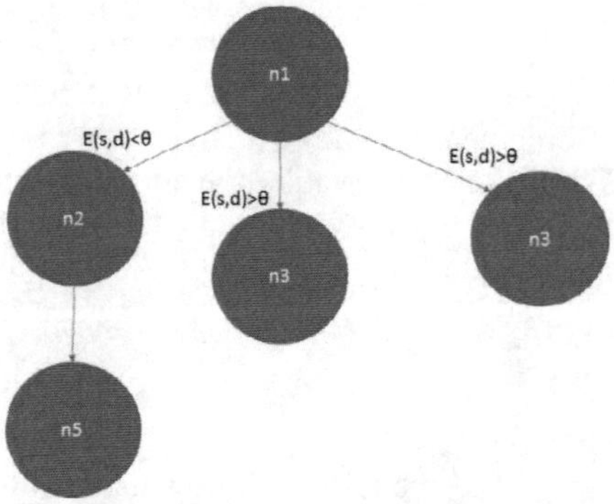

Figure 1. Selection Process of Deeper Search

IOP Conf. Series: Materials Science and Engineering **946** (2020) 012003 doi:10.1088/1757-899X/946/1/012003

4. Experimentation and Result

4.1. *Experimental Setup*

HGM and its rival techniques are simulated using Windows 10 Enterprise with 16GB RAM and dual Intel Core (TM) i7-4702MQ CPUs each of 2.20GHz speed. All experiments are conducted by using the benchmark malwares. The data malware forensic is analysed using different tools for deep visibility of malware behaviours. Wireshark and Network Miner are run on Security Onion to analyse malware files. These data files are then pre-processed to generate a training dataset for the models. The benchmark malware datasets which have been applied for model training are: Microsoft-Malware dataset [18], ARP attack-dataset [19], and ICMP attack-dataset [19]. HGM performance is compared with state-of-the-art supervise learning techniques which are: J48, Convolutional Neural Networks (CNN), Random Forest and Feedforward Neural Networks (FNN). Performance of these models is measured by using three parameters which are Correlation Coefficient (CC), Accuracy and Root Mean Square Error (RMSE). The results of each model are run ten time and then average of these runs is used for comparison purpose.

4.2. *Result*

HGM is compared with its rivals with respect to all three performance measurement parameters. Table 1 shows the best performance of each model.

Table 1. Comparisons of AI models

Model	CC	Accuracy	RMSE
HGM	0.31	0.78	0.038
CNN	0.44	0.74	0.024
J48	0.04	0.49	0.493
Random Forest	0.54	0.64	0.395
FNN	0.088	0.71	0.175

The results show that HGM performs comparatively better than its rivals, particularly in accuracy; however, due to variations in the number of samples in each run, the error rates fluctuate tremendously in each episode of experimental runs. The use of a convolutional neural network in generalization attack behaviors shows a promising strategy. CNN requires extensive training episodes as compared to HGM. Random Forests performs better in terms of correlation coefficient thank other models; however, it does not produce consistent results in terms of accuracy as it does build ensemble models of decision tree and chooses the best one. For this reason, it performs better than J48. HGM is better than its predecessor and other models due to its adaptive approach to sample selection mechanism. The samples are selected based on the quality of the search for solutions during Monte-Carlo simulations.

5. Conclusion

A new approach for malware detection so-called HGM was presented. It is a Monte-Carlo simulation-based model that uses the best features of RBCM and Random Forest. Extensive experimentations were conducted using different malware, known as Microsoft malware, ARP attack, and ICMP attack. HGM demonstrated a better performance than its rival in detecting the malware of various kinds. It demonstrated high performance and self-adaptability to exploration and then balancing the trade-off between exploration and exploitation. Furthermore, as android is ubiquitous and has a very largest share of the mobile OS market with billions of application downloads from the official app market; therefore, we plan to explore HGM for application Malware.

Acknowledgments
This research was funded by the Research Funds of Higher Colleges of Technology (UAE) [HCT-RG-1394]. The authors would like to thank the HCT research office for their support, and also the HCT students who greatly assisted them in this research.

References
[1] Y. Ye, T. Li, D. Adjeroh, and S. S. Iyengar, "A survey on malware detection using data mining techniques," *ACM Comput. Surv.*, vol. 50, no. 3, 2017, doi: 10.1145/3073559.
[2] K. O. Babaagba and S. O. Adesanya, "A study on the effect of feature selection on malware analysis using machine learning," *ACM Int. Conf. Proceeding Ser.*, vol. Part F1481, pp. 51–55, 2019, doi: 10.1145/3318396.3318448.
[3] M. Alrammal, M. Naveed, and S. Rihawi, "Using heuristic approach to build Anti-malware," in *ITT 2018 - Information Technology Trends: Emerging Technologies for Artificial Intelligence*, 2019, doi: 10.1109/CTIT.2018.8649499.
[4] B. J. Kwon, J. Mondal, J. Jang, L. Bilge, and T. Dumitraundefined, "The Dropper Effect: Insights into Malware Distribution with Downloader Graph Analytics," in *Proceedings of the 22nd ACM SIGSAC Conference on Computer and Communications Security*, 2015, pp. 1118–1129, doi: 10.1145/2810103.2813724.
[5] W. Mao, Z. Cai, D. Towsley, and X. Guan, "Probabilistic Inference on Integrity for Access Behavior Based Malware Detection BT - Research in Attacks, Intrusions, and Defenses," 2015, pp. 155–176.
[6] P. Vadrevu, B. Rahbarinia, R. Perdisci, K. Li, and M. Antonakakis, "Measuring and Detecting Malware Downloads in Live Network Traffic BT - Computer Security – ESORICS 2013," 2013, pp. 556–573.
[7] T. Dube, R. Raines, G. Peterson, K. Bauer, M. Grimaila, and S. Rogers, "Malware target recognition via static heuristics," *Comput. Secur.*, vol. 31, no. 1, pp. 137–147, 2012, doi: https://doi.org/10.1016/j.cose.2011.09.002.
[8] Q. K. Ali Mirza, I. Awan, and M. Younas, "CloudIntell: An intelligent malware detection system," *Futur. Gener. Comput. Syst.*, vol. 86, pp. 1042–1053, 2018, doi: https://doi.org/10.1016/j.future.2017.07.016.
[9] E. de O. Andrade, J. Viterbo, C. N. Vasconcelos, J. Guérin, and F. C. Bernardini, "A Model Based on LSTM Neural Networks to Identify Five Different Types of Malware," *Procedia Comput. Sci.*, vol. 159, pp. 182–191, 2019, doi: https://doi.org/10.1016/j.procs.2019.09.173.
[10] J. Saxe and K. Berlin, "Deep neural network based malware detection using two dimensional binary program features," in *2015 10th International Conference on Malicious and Unwanted Software (MALWARE)*, 2015, pp. 11–20, doi: 10.1109/MALWARE.2015.7413680.
[11] I. Firdausi, C. lim, A. Erwin, and A. S. Nugroho, "Analysis of Machine learning Techniques Used in Behavior-Based Malware Detection," in *2010 Second International Conference on Advances in Computing, Control, and Telecommunication Technologies*, 2010, pp. 201–203, doi: 10.1109/ACT.2010.33.
[12] E. Menahem, A. Shabtai, L. Rokach, and Y. Elovici, "Improving malware detection by applying multi-inducer ensemble," *Comput. Stat. Data Anal.*, vol. 53, no. 4, pp. 1483–1494, 2009, doi: https://doi.org/10.1016/j.csda.2008.10.015.
[13] I. Santos, J. Devesa, F. Brezo, J. Nieves, and P. G. Bringas, "OPEM: A Static-Dynamic Approach for Machine-Learning-Based Malware Detection BT - International Joint Conference CISIS'12-ICEUTE´12-SOCO´12 Special Sessions," 2013, pp. 271–280.
[14] M. Naveed and M. Alrammal, "Reinforcement learning model for classification of Youtube movie," *J. Eng. Appl. Sci.*, vol. 12, no. Specialissue9, 2017, doi: 10.3923/jeasci.2017.8746.8750.
[15] M. Naveed, A. Crampton, D. Kitchin, and L. McCluskey, "Real-Time Path Planning using a Simulation-Based Markov Decision Process," in *Research and Development in Intelligent*

Systems XXVIII, 2011, pp. 35–48.

[16] D. Ucci, L. Aniello, and R. Baldoni, "Survey of machine learning techniques for malware analysis," *Comput. Secur.*, vol. 81, pp. 123–147, 2019, doi: https://doi.org/10.1016/j.cose.2018.11.001.

[17] E. M. Karanja, S. Masupe, and M. G. Jeffrey, "Analysis of internet of things malware using image texture features and machine learning techniques," *Internet of Things*, vol. 9, p. 100153, 2020, doi: https://doi.org/10.1016/j.iot.2019.100153.

[18] R. Ronen, M. Radu, C. Feuerstein, E. Yom-Tov, and M. Ahmadi, "Microsoft Malware Classification Challenge." 2018, doi: 10.13140/RG.2.2.34695.91045.

[19] M. Parkour, "Contagio Dataset." http://contagiodump.blogspot.com/ (accessed Feb. 26, 2020).

IOP Conf. Series: Materials Science and Engineering **946** (2020) 012004 doi:10.1088/1757-899X/946/1/012004

Interpretation of the category of "complex" in terms of dialectical positivism

Yelizaveta S Vitulyova[1], Akhat S Bakirov[1,2*], Saltanat T Baipakbayeva[3] and Ibragim E Suleimenov[2]

[1] Almaty University of Power Engineering and Telecommunications, Almaty, Republic of Kazakhstan

[2] National Academy of Engineering of the Republic of Kazakhstan, Almaty, Republic of Kazakhstan

[3] Institute of Information and Computational Technologies, Almaty, Republic of Kazakhstan

axatmr@mail.ru

Abstract. The interpretation of the category of complex is given on the basis of dialectical positivism, in the framework of which the principle of dialectic symmetry is formulated. It, in turn, is based on the consideration of matter and information as paired dialectic categories. In accordance with this principle, information objects form a hierarchy similar to that which corresponds to different levels of matter organization (mechanical, chemical, biological, social). As "complex", one should interpret the system that is the medium for the information object at least one level higher than the level related to trivial information (that is, one that is inextricably linked with the object itself). This level is the level corresponding to estranged information (simplifying information about a certain object recorded on another medium).

1. Introduction

Human civilization is currently dealing with increasingly complex systems. This, to put it mildly, actualizes the analysis of the concept of "complex", or rather, the analysis of the category of "complex", understood in the philosophical meaning of this term. It would not be an exaggeration to say that science, once created during the Enlightenment, was an answer to the challenges that civilization faced on the part of the forces of nature. However, in its development, civilization has generated numerous new challenges, and all of them are somehow connected with the appearance of more and more complex systems created by man himself.

The largest thinkers of the 20th century have repeatedly made attempts to understand the nature of the complex as such. Here, one cannot fail to emphasize the systematic approach of von Bertalanffy, the ideas of Ilya Prigozhin and much more. In modern literature, numerous systems are also analyzed that satisfy, in one sense or another, the criterion of "complexity", in particular, [1-4]. These and other works clearly demonstrate that very often when considering "complex" systems, the nature of their constituent elements becomes secondary. The structure of relations between the elements of the system comes to the fore.

Moreover, many of the works performed in the same vein as those cited above, for example, [5,6], consider extremely abstract complex systems in which elements are structureless, and their only property is the ability to form certain connections with other elements of the system. All these works are certainly more than serious interest, and the results obtained in them already allow us to raise the question of the corresponding generalizations of the philosophical level.

Content from this work may be used under the terms of the Creative Commons Attribution 3.0 licence. Any further distribution of this work must maintain attribution to the author(s) and the title of the work, journal citation and DOI.
Published under licence by IOP Publishing Ltd

In this paper, on the basis of the philosophy of dialectical positivism [7-9], an attempt is made to prove that the nature of the complex is purely informational. In other words, a system becomes complex when the processes of exchanging information between its components under the systems come to the fore, more precisely, a complex system cannot but be a system for processing information, albeit of the simplest type.

We emphasize that it is precisely this view of the nature of complex that is also being updated in connection with the problems of artificial intelligence. Indeed, any form of human intellectual activity is somehow related to the existence of a very specific complex system - the brain, and they appear as a result of the exchange of signals between relatively simple elements - neurons. Each individual neuron that is part of the brain is able to perform only relatively simple functions. However, together they give rise to some new quality - human consciousness, mind, intellect and so on.

This example can be generalized. Namely, we argue that the exchange of information between relatively simple elements of the system under certain conditions can give rise to a new quality, that is, some information objects that have relative independence, just as the exchange of information between brain neurons generates consciousness. This is the most obvious example of an informational object with relative independence. Consciousness depends on the functioning of all neurons of the brain in the aggregate, but very weakly depends on each of them individually. At a minimum, this is a system property.

Another example of a relatively independent information object is any particular science - its carrier is the entire scientific community as a whole, this information object is preserved even when individual scientists pass away. Moreover, modern science of science claims [10] that very often specific scientific areas develop, subject to their own logic of development, i.e. the degree of independence of this object from system elements (individual researchers) is much greater than it might seem at first glance.

An example of the human brain in this regard, of course, is partly debatable. Namely, it is clear in advance that brain neurons are a system designed to process signals. It is not obvious that elements of complex systems of a different nature can also be considered in terms of information processing processes. Based on this, it is not obvious that this example can be generalized to complex systems of arbitrary nature.

However, for all the debatability, this example nevertheless unequivocally shows that in a complex system (at least of a certain type), under certain conditions, a new quality may appear - a relatively independent information object, only indirectly associated with the elements that make up the system.

This is a new quality that does not appear due to the fact that elements have certain properties, but because they have the ability to form relationships with each other. There are conditions when the nature of the relationship between the elements of the system undergoes a qualitatively-quantitative transition (this term is used in the sense that classical dialectics give it), as a result of which something new is born in the system - an object that cannot have any other nature than information . In any case, its consistent analysis can be carried out in terms of information theory, since the properties of the elements themselves that make up the system remain unchanged.

It is in this sense that in this work any systems that satisfy the criterion of complexity are treated. A new - informational - quality appears in them.

The proof of the statements made is given in this paper from the standpoint of dialectical positivism [7-9]. Examples related to the field of specific sciences are purely illustrative.

2. Informational nature of the complex: justification based on the principle of dialectical symmetry

In accordance with the point of view of dialectical positivism [7-9], information should be considered as a dialectical category, a paired category of matter. As you know, objective dialectics defines the basic concepts called categories through opposition. It is impossible to give them the usual definition, understood in the "school" sense of the word precisely because they are the most fundamental, and an attempt to reveal the meaning of these terms through others inevitably leads to a vicious logical circle.

Once such a definition of the category of information is given, the next step can be taken immediately. Namely, there is a well-defined hierarchy of levels of organization of matter; there is a level associated with the mechanical movement, then with the chemical, biological and, finally, social. (All these levels of organization of matter, of course, have sublevels.)

The principle of dialectic symmetry [9], starting from the above definition of information given by the philosophy of dialectical positivism, argues that a similar hierarchy exists in relation to all those entities that are associated with the category of information.

To simplify somewhat, we can say this: a hierarchy of levels of organization exists in the "material world", and accordingly a similar hierarchy must exist in the "information world", which directly follows from the law of unity and struggle of Hegel's opposites. Of course, the concepts of “material world” and

IOP Conf. Series: Materials Science and Engineering **946** (2020) 012004 doi:10.1088/1757-899X/946/1/012004

"information world" in the statement stated above should be understood rather metaphorically, since "material" and "information" are inextricably linked with each other.

Consider a hierarchy of information objects. The simplest form of information is that associated with individual messages. This is, for example, a binary number record or a record made by the letters of another alphabet. At the next level of the hierarchy of information entities are information objects, which are the rules for operating with information of the simplest type, in particular, the rules for operating with binary numbers.

Indeed, the arithmetic rules of addition, multiplication, etc. it is also information, but it is information of a special type. It allows you to work with other information objects (in particular, add binary numbers) and thereby obtain additional information (the result of addition, for example).

At an even higher level are information objects that acquire relative independence in relation to their medium. An example is mathematics in general. This is a well-defined information object that is capable of developing within the framework of its own logic.

Yes, of course, this development is carried out by mathematicians, but the result of their collective efforts is not connected with any of the mathematicians individually, it is connected with the entire mathematical community as a whole and develops, subject to its own laws. It would not be an exaggeration to say that mathematics as an information object is registered in the noosphere (understood in the sense that V.I. Vernadsky attached to this term [10]) as a whole. Simplifying somewhat - mathematics is much more than the sum of knowledge that each mathematician has individually.

Let us return to the hierarchy of information objects.

Obviously, here it is necessary to distinguish some lower level of the hierarchy, which separates just information (for example, messages) from more complex information objects. From the point of view of dialectical positivism, this level is associated with objects capable of, at a minimum, capturing certain information. The recorded information is the simplest information object that allows you to interpret the category of complex from the proposed point of view. Let us explain the above with an example.

Each of us is witnessing a wide variety of events. For example, it may simply be that a passenger forgot a book in the subway or made some other minor action. It is obvious that in such events, as in any others: the material and informational aspects are dialectically inextricably linked with each other. By studying individual water samples, one can obtain the necessary information about the physical and chemical properties of water in general.

Returning to the example of insignificant events, we can say this: this event either goes unnoticed, or someone fixes it in his memory (or writes it in a diary), or carries out some other actions, as a result of which this information can be replicated further.

Considerations of this kind lead to ideas about alienated information [8.9]. The fact of recording information about an event indicates that the information is alienated here. Likewise, the scientific knowledge obtained in the study of samples of a chemical substance also leads to the fact that information is *alienated* from the original objects. In a certain sense, the dialectical unity between the matter of information, originally inherent in these objects themselves, is being destroyed.

So, fixing information, transferring it to another medium also corresponds to the lowest level in the hierarchy of information objects.

At the next level of the hierarchy are information objects that form certain *information processing systems*. So, the rules for operating with binary or decimal numbers also represent information (it can be fixed, for example, recorded in a textbook). But, this information is something qualitatively different in comparison with a "simple" message - the rules for operating with binary or decimal numbers allow you to receive new information, for example, to calculate the characteristics of a non-existing electronic circuit.

At an even higher level of the proposed hierarchy are information objects, which are information processing systems capable of generating other information processing systems related to a lower hierarchical level. An example of such an object is geometry as a whole; it is a means of obtaining theorems that can themselves be used to generate new information (for example, when performing calculations). Continuing this logic, we can come to the following interpretation of the concept of intelligence; it is such an information processing system that is capable of generating information processing systems of such a high level that they allow us to make judgments about the original system. In the language of the humanities, this means - such a system is able to recognize itself.

Thus, the principle of dialectical symmetry allows you to enter into use a certain hierarchy of information objects, and accordingly - a hierarchy of systems of an arbitrary nature in terms of "complexity". Any information object is dialectically connected with its carrier; therefore, by a "complex"

IOP Conf. Series: Materials Science and Engineering **946** (2020) 012004 doi:10.1088/1757-899X/946/1/012004

system of a certain hierarchy level, one can understand that which is a carrier for an information object of a corresponding complexity level.

As a "complex" system (without additional refinements), it is therefore logical to interpret one that is capable of generating information objects, at least one level higher than the minimum. Therefore, these are systems in which processes of alienation of information can take place.

It is appropriate to emphasize here that there are a huge number of complex systems in which we can talk about the alienation of information. For example, in the chemistry of macromolecular compounds, the so-called customizable sorbents are widely known [11]. The configuration of the polymer is selected so that it selectively captures the atoms or molecules of a certain substance, which also have the corresponding geometric characteristics. Obviously, a customizable sorbent can also be interpreted from an information point of view. This is a connection that stores information about another connection. Biological informational macromolecules - RNAs - perform similar functions, but obviously related to a higher level of "complexity" in the above sense. This example can be generalized. In particular, returning to the works cited above, in which complex systems of various nature (as well as systems formed by abstract "elements") were studied, it can be noted that many of them are characterized by so-called "phase transitions" - the system jumps from one state to another. Moreover, as shown, for example, in [12], this transition is often accompanied by hysteresis phenomena.

Obviously, if we are talking about a complex system, then it is always possible to identify well-defined local fragments in it, and, therefore, any distributed complex system, some parts of which can undergo a phase transition relatively independently of others, can be considered as an analogue of a neural network[13].

Indeed, any relatively independent fragment of such a system can be associated with a neuron. The absence of a phase transition is treated as a logical zero, the fact of a phase transition is treated as a logical unit. As long as the system is still a whole, the phase transitions in individual parts of the system will inevitably affect at least the neighboring ones (Fig. 1). In accordance with the scheme shown in this figure, a complex system is conditionally divided into regions, in each of which a phase transition occurs. Arrows indicate the effect of regions on each other. In other words, a feedback system arises here, and the system as a whole can be considered as an analog of the Hopfield neuroprocessor (Fig. 2). An example of a specific system in which nontrivial phase transitions corresponding to the model of Fig. 1 are realized is considered in [14].

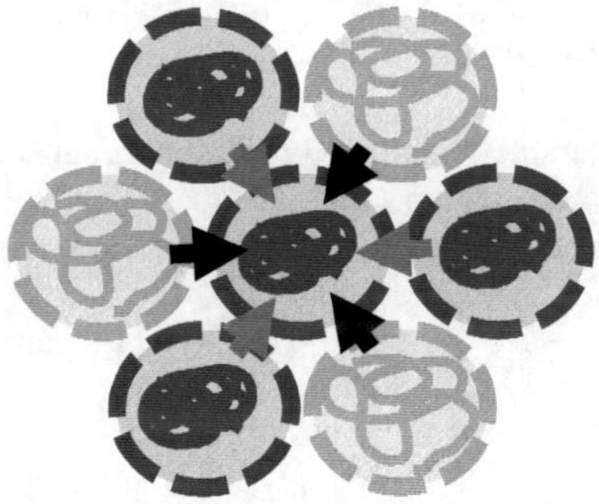

Figure 1. Inhomogeneous system in which phase transitions take place, as an analog of a neural network

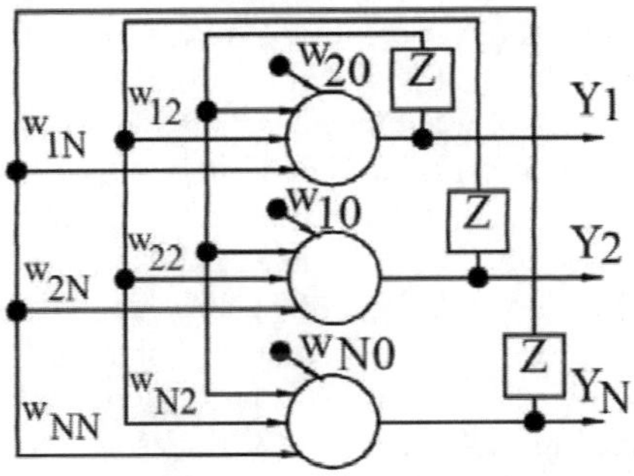

Figure 2. Schematic of the Hopfield neuroprocessor

Accordingly, any external influence on such a system can be considered in the language of information theory: what leads to phase transitions can be described in terms of logical units and zeros (which becomes especially clear when the phase transition leads to the appearance of inhomogeneities).

This example also allows generalization. Namely, it shows that a system that has crossed a certain threshold of complexity, at a minimum, can be considered as something where information reading processes take place. Moreover, since there is a certain interaction between the parts of the system, we can say that the system of the considered level of complexity provides the reading of information from one part with the help of another. If this condition is met, then an information object of the lowest hierarchy level, which was mentioned above, is obviously formed in it. Moreover, this is a prerequisite for the formation of information objects related to higher levels of the hierarchy.

3. Conclusion

Thus, the principle of dialectical symmetry allows us to give the following interpretation of the category of "complex". A system is treated as complex when information objects are formed in it that lie at least one level higher than trivial (inalienable information), or prerequisites are created for the appearance of such objects.

The lowest level in the hierarchy of complex systems is occupied by those that provide alienation (reading information), i.e. By studying them, one can obtain information not only about these systems themselves, but also about some others.

At higher levels of the hierarchy of complex systems are systems that are carriers for information processing systems.

4. References

[1] Wang X F 2002 Complex networks: topology, dynamics and synchronization *International journal of bifurcation and chaos* **12(5)** pp 885-916 DOI: https://doi.org/10.1142/S0218127402004802
[2] Albert R, Barabási A L 2002 Statistical mechanics of complex networks *Reviews of modern physics* **74(1)** p 47 DOI:https://doi.org/10.1103/RevModPhys.74.47
[3] Gross T, Blasius B 2008 Adaptive coevolutionary networks: a review *J. R. Soc. Interface* **5(20)** pp 259–71 DOI: https://doi.org/10.1098/rsif.2007.1229
[4] Suleimenov I and Bakirov A 2018 Semi-empirical Theory of Distributions of Users of Social Online Networks by the Number of Contacts *MATEC Web Conf.* **214** DOI: 10.1051/matecconf/201821402001
[5] Solé R V, Manrubia S C, Luque B, Delgado J, Bascompte J 1996 Phase transitions and complex systems: Simple, nonlinear models capture complex systems at the edge of chaos *Complexity* **1(4)** pp 13-26 DOI: https://doi.org/10.1002/cplx.6130010405
[6] Radin C, Sadun L 2013 Phase transitions in a complex network *Journal of Physics A: Mathematical and Theoretical* **46(30)** DOI: https://doi.org/10.1088/1751-8113/46/30/305002
[7] Suleimenov I, Massalimova A, Bakirov A and Gabrielyan O 2018 Neural Networks and the Philosophy of Dialectical Positivism *MATEC Web Conf.* **214** DOI: 10.1051/matecconf/201821402002
[8] Suleimenov I E, Gabrielyan O A, Bakirov A S and Vitulyova Ye S 2019 Dialectical Understanding of Information in the Context of the Artificial Intelligence Problems *IOP Conference Series: Materials Science and Engineering* **630** DOI: 10.1088/1757-899X/630/1/012007
[9] Gabrielyan O, Suleimenov I, Vituleva E 2019 Artificial intelligence in the context of noosphere studies SCTCMG 2019 International Scientific Conference «Social and Cultural Transformations in the Context of Modern Globalism» https://doi.org/10.15405/epsbs.2019.12.04.130
[10] Vernadsky V I 2013 A few words about the noosphere *Noosphere studies* **1** pp 6-17
[11] Belka M, Ulenberg S, Bączek T 2017 Fused deposition modeling enables the low-cost fabrication of porous, customized-shape sorbents for small-molecule extraction *Analytical chemistry* **89(8)** pp 4373-76 DOI: https://doi.org/10.1021/acs.analchem.6b04390
[12] Suleimenov I E, Guven O, Mun G A, Uzun C, Gabrielyan O A, Kabdushev S B, Nurtazin A 2017 Hysteresis effects during the phase transition in solutions of temperature sensitive polymers *Eurasian Chemico-Technological Journal* **19(1)** pp 41-46 DOI: https://doi.org/10.18321/ectj501
[13] Ibragim S. et al. 2020 Example of the use of artificial neural network in the educational process *Future of Information and Communication Conference, Springer, Cham* pp 420-430 DOI: https://doi.org/10.1007/978-3-030-39445-5_31

[14] Suleimenov I, Güven O, Mun G, Beissegul A, Panchenko S, Ivlev R 2013 The formation of interpolymer complexes and hydrophilic associates of poly (acrylic acid) and non-ionic copolymers based on 2-hydroxyethylacrylate in aqueous solutions *Polymer International* **62(9)** pp 1310-15 DOI: https://doi.org/10.1002/pi.4422

IOP Conf. Series: Materials Science and Engineering **946** (2020) 012005 doi:10.1088/1757-899X/946/1/012005

Improving Aviation Incidents using Association Rule Mining Algorithm and Time Series Analysis

Pierre Pauline R. Abesamis, Remedios de Dios Bulos, and Michelle Ching

De La Salle University, 2401 Taft Avenue, Manila, Philippines

pierre.abesamis@dlsu.edu.ph; remedios.bulos@dlsu.edu.ph;
michelle.ching@dlsu.edu.ph

Abstract. Advance technology helps to forecast and show the different hazards associated in the world of transportation systems. Despite the rapid advancement of technology, there are still huge number of aviation accidents and incidents happened. This paper concerns application of various data analytics algorithm to derive a time series forecasting method and frequent accident and incident pattern specifically for Philippine aviation setting for predicting conditions that would increase the likelihood of aviation accidents and incidents involving fatal and non-fatal activities. Data Analytics algorithms are Association Rule Mining using FP-Growth algorithms and Time Series Forecasting methods using Linear Regression, Gaussian Processes, Multilayer Perceptron, and SMOreg are applied to datasets derived from original data obtained from Department of Transportation (DOTr) with its attached agency, Civil Aviation Authority of the Philippines (CAAP) and both incident and accident records stored in *"Philippine Aviation Incident Reporting System (PAIRS)"* from 2008-2017. The aviation accident data is based on flight information such as aircraft attributes, aviation accident factors, type of occurrences, geographical location, weather conditions and phases of operation. The results lead to prescriptive analytics for development of business rules and help identify pattern that can define aircraft accidents/incidents occurrences. The ability to predict accidents has the main objective of saving lives and will be impactful in cost saving in terms of any aircraft damages. The result created important points in improving the aviation operation and assists in better decision-making to help create policy and future improvement for aviation-safety in the country.

Index Terms— *Analytics Value Chain, Aviation Accidents, Association Rule Mining Algorithm, Time Series Analysis*

1. Introduction

1.1. Background of the Study

There have been several improvements in safety during flight that led to increase the attention to on-ground risks in the aviation industry that can occur before take-off and after landing. Improvement for the aviation safety has been defined over the years. Most of them are related to technological improvements in aircrafts, avionics and engines that resulted to better aviation safety record [1]. One study according to National Airspace System (NAS) has identified the value of unknown potential hazards and associated risks in improving the system safety. It used data analytics algorithm to create a risk model to identify causal factors [2]. The other is the pilot training. The pilot training has

Content from this work may be used under the terms of the Creative Commons Attribution 3.0 licence. Any further distribution of this work must maintain attribution to the author(s) and the title of the work, journal citation and DOI.
Published under licence by IOP Publishing Ltd

IOP Conf. Series: Materials Science and Engineering **946** (2020) 012005 doi:10.1088/1757-899X/946/1/012005

benefited the improved understanding of human factors and the application of training and regulations. Also, both navigational aids and air traffic management has improved to initiate safer flights with improved weather forecast. In addition, one of the major contributions is improving the safety record that can be traced to thorough investigation within accident and incidents past records which helps prevent for such events to occur. Hence, this reactive approach improves the aviation safety and enhances the data analysis within the various records for both incidents and accidents related to aviation [1].

The aviation data analysis needs to have analysis methods that are effective at finding interesting patterns in the collected data. The analysis of data relies on the past experiences within the aviation accidents and incidents that are caused by multiple factors. Analyzing both accident and incident data can show relationships associated between them. Aviation collection can be from a combination of automated and manual method. However, in this study it will only cater the automated reported events using the *PAIR* system.

In the Philippines, this is under the Department of Transportation (DOTr) agency with the cooperation of Civil Aviation Authority of the Philippines (CAAP) on which they handle all the aviation accident and incidents. Currently, they have implemented the *"Philippine Aviation Incident Reporting System (PAIRS)"* program [3]. Its objective is to promote the accident prevention by analysis of safety data. The reporting system facilitates the collection of actual and potential safety deficiencies that include both accidents and incidents report. Also, *PAIRS* focus on systems, human factors, procedures and equipment and covers all occurrence on which any event can happen. This paper will study the associated factors of frequent incidents in line with the goal of accident prevention. It can help associate rules of frequent aviation accidents to improve the aviation safety and help the aviation community for recommendation for reviewing policy and other plans for improvements.

In this paper, it will use the factors and occurrence for both incident and accident factors based on the report extracted by CAAP in *PAIRS* system that would be used as data set for this study. It will be tested from the year of 2008 to 2017 that will cover all incident and accident reported. The data used for this research study were collected from Open Data Philippines. Open Data Philippines is an online platform that allows government agencies to be transparent with their actual data and be visible to its citizens. Despite seeing the results of the collected data, DOTr still have unconsolidated data and needs data integration with its different attached agencies such as in CAAP to analyze the various data sets. Thus, today there is a need to learn from past data to apply prediction of future occurrences of events using the past data. Overall, this study can help them decide for the process improvement and policy enhancement for the enhancement of aviation-safety.

1.2. Objectives of the Study

This research study aims to help CAAP and DOTr in consolidating and analyzing data in the past years for both incident and accident aviation events by having quantitative data analytics. The main objective of this paper is to identify associated factors of frequent incidents in line with the goal of accident prevention. Furthermore, it will identify significant accident factors and evaluate the trends of aircraft accident and incident events reported in the years of 2008 to 2017. This information is helpful for the agency to lessen frequency of accidents and to maximize the contributing variables in promoting aviation accident safety.

It will be done using Association Rule Mining using the FP-Growth algorithm in RapidMiner data science tool to formulate associating rules of pattern for aviation incidents that covers both incidents and accidents. Moreover, time series forecasting using Waikato Environment for Knowledge Analysis (Weka) machine learning workbench will be used to conduct a one-year predicted data and result with be compared to determine the most accurate prediction using Linear Regression, Gaussian Processes, Multilayer Perceptron, and SMO Regression. This is useful to understand its pattern to extract valuable information that can help recommend aviation-safety strategies.

1.3. Scope and Limitation of the Study

Open Data Philippines has different kinds of aviation data sets provided by CAAP as attached agency of DOTr that handles all the incidents for both aviation incidents and accidents. The focus of this study will only for reported aircraft incidents and accidents using the *Philippine Aviation Incident Reporting System (PAIR) program* within the years of 2008-2017. Initially, the data sets will only cover the standard aircraft incident and accident reports and will not be covering the data that were not reported and not retrieved in the system. Also, it will cover all location within the Philippines both in cities and provinces and all of the incident and accident factors would be considered in the study.

Moreover, the researcher chose RapidMiner as the data science tool to be used for conducting the Association Rule Mining using the Frequent Pattern (FP-Growth) Algorithm to find association rules of frequent aviation accident pattern or items sets mining in market basket analysis and to be able to make variations [4]. Another tool is Weka's Forecast package will be used for time series forecasting using Linear Regression, Gaussian Processes, Multilayer Perceptron, and SMO Regression to determine the most accurate one-year forecast for the identified *PAIRS* factors (human, equipment, procedures and systems). In addition, it will show the descriptive part using the MS Excel 2016's Analysis ToolPak that allows data analysis tools to have statistical and engineering analysis in data visualizations and descriptive analysis. It is used to provide the visual interface on a workflow and user-friendly environment. It will also be integrated within R programming language that shows a complete visual workbench in delivery appropriate result and report. Aside from the above tools, MS Excel 2016's Analysis ToolPak together with R programming language. It allows data analysis tools to have statistical and engineering analysis in data visualizations and descriptive analytics.

2. Review of Related Literature Work

2.1. Related Research Studies

Data Analysis in accident is very helpful since it can reveal the relationship between the different types of attributes that can lead to an accident. Several types of road accidents such as in road, traffic and airplane have different ways of comparison and nature since they are all unpredictable. However, this study will only focus on aviation accidents.

Data mining is a process on which it will help create useful and understandable knowledge within the data [5]. Thus, it should be used to examine the aviation accident data set and particularly it will focus on Association Rule Mining and Time Series Forecasting [6]. Based on literature, Aviation safety data analysis commonly face issues of finding analysis methods that are effective at finding interesting patterns within the collected data [7].

2.2 Association Rule Mining

Association Rules Mining is one of the most common and very important aspect of data mining for it is being used to find some interesting associations or patterns within correlation relationships between item sets within a huge amount of data [8]. It is commonly used to identify frequent item sets and is one of the most powerful tools for technology and a useful step in the application of association rules mining [9]. Also, it is a technique on which it patterns to the market basket analysis of market data and focus is on identifying the goods on which the customer prefers to buy together more frequently. Therefore, it produces several sets of rules that can improve the correlation among different set of attributes in data set. It has interesting measures like the *support* and *confidence* that results to have stronger rules. The support value shows the frequency of occurrence of a rule on a data set and the confidence illustrates about the reliability of a rule. Thus, a high confidence and support value is the main interest on any association rules. In addition, clusters analyze using association rule mining find the correlation with data attributes. And each cluster is analyze using the association rule mining algorithm like FP-Growth (frequent-pattern growth) algorithm [10].

FP-Growth algorithm is the improved version of the Apriori algorithm that has been started by Jiawei Han. It compresses the data sets to a FP-tree (frequent-pattern tree) and scans the database twice but still maintains the information of associations between item sets. The compressed database is divided into sets of condition database and each condition associates with a frequent item. Also, it does not produce a candidate item sets in mining process, but it greatly improves its mining efficiency.

However, the FP-Growth algorithm needs to have a FP-tree that contains all data sets which occupies higher requirement in memory space [5].

2.3. Time Series Forecasting

Time series method is being used to analyze any statistical characteristics of aviation accident data. In most statistical data, there are logical events that can shows the periodicity and timing that affects the statistics by variety of factors. Time series is based from past historical data that is shown in a regular basis that is combined together with influencing factors of data to help make future predictions of data. In this paper, it is an important feature that contributes to the significance of aviation accidents trend analysis [11].

Time series forecasting analysis composed of description of the component's movements of the present data. There is a graphical representation for example: time plot reveals quantitatively the presence of long-term trend, cyclical, seasonal and irregular variation. Time series analysis also investigates the following factors and in decomposition of time series into basic component movements. The time plot enables to construct the long trend term curve that can be used as (a) method of freehand, (b) least squares methods, (c) moving average method and (d) semi-average method [12].

For this study, the research was examined and applied in various forecasting techniques used in Weka. Weka is a tool to compare various prediction functions and found that SMOreg (SMO regression) function showed the ability for the most realistic and accurate aviation accident prediction other than the existing methods such as the Linear Regression, Multilayer Perceptron and Gaussian processes. Some of the researchers also recommended to use either SMOreg or Gaussian processes to determine the most accurate prediction.

3. Methodology

In this section, it explained how the collection and preparation of aviation accident dataset were done. It provides a discussion on how the data were analyzed and used for statistical analysis. Also, it includes how data mining models were created to answer the objectives of this research. This study will use various tools namely: (a) RapidMiner as data analytics tool to demonstrate Association Rules Mining particularly the FP-growth algorithm, (b) Weka's Forecasting package for the Time Series forecasting using Linear Regression, Gaussian Processes, Multilayer Perceptron, and SMOreg, and (c) MS Excel 2016's Analysis ToolPak to show the descriptive analytics, and visualizations. The below section discusses the study's operational framework and algorithm used for the whole study.

3.1. Operational Framework

The Analytics Value Chain as shown in Figure 1 is the operational framework used for this research study.

Figure 1. Analytics value chain.

In the Analytics Value Chain, it begins with the curate phase on which it covers the process of collection, storage, management and cleaning of data. The second phase is the summarize for it includes the charts, tables and visualization techniques to summarize the clean version data sets. In the descriptive part, it is used to know the past and present trends while the predictive focuses more on the forecasting outcomes of possible events. Those data can be helpful upon creating possible business rules in the future. The prescriptive shows recommendation of possible actions to do based on all the data gathered as discussed by IBM Smarter Analytics (2013).

3.2 Data Collection

To test and evaluate the aviation accident prediction and forecast models, the collected historical accident and incident data were gathered from the Civil Aviation Authority of the Philippines (CAAP) on which it is an attached agency of Department of Transportation (DOTr) that handles all the aircraft issues and aviation accidents. With the help of the *"Philippine Aviation Incident Reporting System (PAIRS)"*, they can extract an excel reports that can be useful collection of information on actual or potential safety deficiencies that is being captured by the mandatory incident reporting system [3].

The aviation accident dataset consists of all the location in the Philippines and within the focus area of the *PAIRS* which are systems, human factors, procedures and equipment. It also supports all the areas covered within the PAIRS program which include departure/in route/approach/landing operations, aircraft ground operations, movement on the airport, fueling operations, safety-related passenger, airport conditions or services or any aviation-safety related issues that were not categorized. The duration of years is from 2008-2017. Initially, a total of 1,214 data were used to create the model. However, all the data set that were used in the study were only gathered in the *Open Data* portal on which it shows the available government data open for public used.

3.3 Data Preprocessing
Data pre-processing is the important step in data mining process. It has the phrase of *"Garbage In, Garbage Out"* concept that is available to data mining and machine learning. Analyzing data that is not cleaned thoroughly produces a lot of problems that can lead to unreliable results. With this, quality of data is being done first before going to an actual analysis of data. Data preparation purpose is filtering steps that can take huge amount of time. The data preprocessing has data cleaning, data integration, data transformation and data reduction. The results of this can now be used for data mining [13].

3.3.1 Data Cleaning
Data would be analyzed using the data mining techniques and at first it can be incomplete such as lack of attributes values, attributes of interest, noisy (consists of errors) and inconsistent (discrepancies of other data) [13]. In this section, there were some missing and inconsistent values within the attributes of factors, weather condition, phase of operation, impact and type of occurrence of the accident. Instances that lacked those attributes were filled in from the dataset. In addition, the duplicate instances were solved. The *date* was separated into (month, day and year) so that the year can be used as separate attribute. The *place of occurrence* was also classified into a province attribute so that some of the redundant and duplicate entries can be grouped into one location. Moreover, some of the *type of occurrence* have new occurrence that is not relevant to the normal group which makes it a noisy data. Also, *report type, status and report attribute* were removed since they are not going to be helpful in the model.

3.3.2 Data Integration
Data integration involves combining of data from multiples sources. In this case, there are separate reports created by the *PAIRS* system. Reports produced were categorized into accidents and incident and year per year generated reports. In this paper it integrates the separate excel sheets that has several reports within accident and incident category and extracted per year basis.

3.3.3 Data Transformation
The data transformation, the data are transformed into consolidated forms needed for data mining. Within the data set, there is the generalization of the data on which the low level or "primitive" (raw) data replaced by higher level concepts through the use of concept hierarchies [13]. In this case, the categorical attributes like the *place of occurrence* and the *type of occurrence* where generalized to higher level concept which are *"province"* and *"group of occurrences."* Hence, it will help within the analyzing data and producing better patterns for aviation accident prevention.

IOP Conf. Series: Materials Science and Engineering **946** (2020) 012005 doi:10.1088/1757-899X/946/1/012005

Table 1. Key aviation accident attributes table.

Attribute Name	Description	Data Type	Value
factors	PAIRS program key indicator	Nominal	Systems, human factors, procedures and equipment
group_of_occurance	The type of occurrence that aircraft accident/incident occur	Nominal	Ditch, Fuel Starvation, Engine Failure, Excursion, Crash, Overshoot, Undershoot, Tail Strike, Gear Failure, UFIT, Stall, Collision, Collapse, Fire, Propeller Strike, Loss Control CFIT, Nature, Abort, Tire Failure, Taxiway Excursion and Others
phase_of_operation	Identifies what operation phase the accident happened	Nominal	Landing, Inflight and Take-Off
weather_condition	Identifies the weather condition the time the accident occurred	Nominal	Visibility, Wind and Cloud
impact	Determines the impact damage of the accident	Nominal	Fatal and non-fatal
province_of_occurence	Shows the location of the accident	Nominal	Cebu, Baclaran, Zambales, Manila, Bulacan, Palawan, Bohol, Davao, Aklan, Pampanga, Iloilo, Las Pinas, Leyte, Quezon City, South Cotabato, Tarlac, Albay, Paranaque, Subic, Baguio City, Pangasinan, Masbate, Antique, Bukidnon, Surigao Del Sur, Kalinga, Pasay City, Samar, Nueva Vizcaya, Mindoro Occidental, Tacloban, Ilocos Sur, Agusan Del Norte, Cavite, Romblon, Tuguegarao, La Union, Bacolod, Dumaguete and Maguindanao
year	The year the accident happened	Nominal	2008-2017

3.3.4 Data Reduction

The concept of data reduction in this study applied the reduction of the dimensions (total number of attributes) [13]. It used the dimension reduction on which it removed the irrelevant, weakly relevant and the redundant attributes were detected and removed. It removed the attributes *status* and *reporting* for it would not be useful in developing the model. Also, it now depends on the discretization and concept hierarchy generation on which the raw data values for attributes are replaced by ranges or higher conceptual levels. These are both done in the type of occurrence and place of occurrence attribute within the dataset. The result of data reduction is presented in Table 1.

3.3.5 Data Mining

Data mining is referred to as knowledge discovery in databases (KDD). It is the process for sorting within huge datasets to identify patterns and establish possible relationships to solve a particular problem in doing data analysis. Also, data mining is being used to predict future trends. The data mining tools is being used to perform data analysis that uncovers important data patterns and

IOP Conf. Series: Materials Science and Engineering **946** (2020) 012005 doi:10.1088/1757-899X/946/1/012005

contributes to help in business strategies [14]. The data mining technique used in the paper is *Find Similar* and is designed to search a collection of aircrafts incidents and accidents to find the most similar to selected accidents. It would be useful in determining if similar incidents or accidents occurred before and if how they are addressed. In the bottom section of the study, it will apply the Association Rules Mining and Time Series forecasting.

4. Data Analytics

Data Analytics is currently being used to describe the analysis of large volume of data on which it shows unique computation tool in handling big amount of data in any organization [15]. It can be both qualitative and quantitative data analysis in ensuring productivity that is achieved using extracting, categorizing and analyzing data in finding meaningful patterns that can help the organization. With this, the researcher aims to study the data set for aviation accident under the *PAIRS* program in the Philippines. It will apply data analytics and will follow the Analytics Value Chain Operational Framework and will be discussed within the paper [16].

4.1 Data Curation and Summary

The focus of the research study is on the aviation incident and accident data sets reported in the year of 2008-2017 that extracted from the *"Philippine Aviation Incident Reporting System (PAIRS)"* program. It includes the attributes *Factors, type of occurrence, phase of operation, weather condition, impact, year and province of occurrence.* Nevertheless, there are duplications with the type of occurrence and province of occurrence which are cleaned and group together to have a standard name for each. The cleaning of data is done manually using MS Excel and resulted to having Table 2 for PAIRS factors and Table 3 for types of occurrence.

Table 2. PAIRS factor Data Set Summary

PAIRS factors	Fatal Accidents	Non-Fatal Accidents
Equipment	144	210
Human factors	120	168
Procedures	96	174
Systems	156	150

Table 3. Types of Occurrence Data Set Summary

Type of Occurrence	Fatal Accidents	Non-Fatal Accidents
Abort	0	6
CFIT	0	6
Collapse	36	12
Collision	36	24
Crash	12	54
Ditch	12	6
Engine Failure	204	174
Excursion	48	84
Fire	6	30
Fuel Starvation	0	18
Gear Failure	60	120
Loss Control	6	6
Nature	6	0
Other	0	42
Overshoot	12	18
Propeller Strike	12	12

Stall	24	12
Tail Strike	6	24
Taxiway Excursion	6	12
Tire Failure	12	12
UFIT	18	6
Undershoot	0	24

4.2 Descriptive Analytics

The Descriptive analytics uses the business intelligence and data mining that answers the question *"What has happened?"* it mines data to provide the trending information on past and current events and can give a real estate of what happened that is in need for future movements. Using the descriptive data accumulated over time, predictive analytics utilizes models over time. To summarize the important data that is related to the aviation accidents and incidents, below are the different tables, figures and visualizations [17].

The collected data shows in Figure 2 presents the total number of aviation accident and incidents within the span of the year 2008 to 2017. It shows that there are big number of non-fatal impacts that total of 58% against the fatal impact of 42%. Within the visualize histogram, there are a lot of accidents that happened in 2013 that cause non-fatal accident while on the year 2011 it shows higher value fatal aviation accidents.

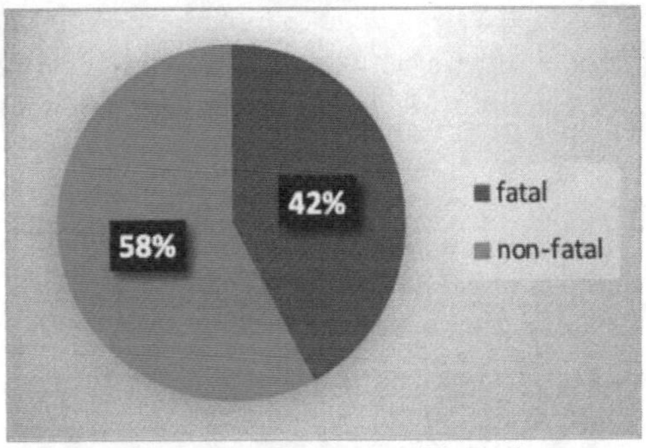

Figure 2. Total aviation accidents impacts.

Another important attribute used in the data is the PAIRS factors. In the below histogram, it shows the comparison of the different PAIRS factors which are the equipment, human factors, procedures and systems that results to both fatal and non-fatal impacts. The diagram in Figure 3 shows that the top *PAIRS factor* is the equipment that leads to non-fatal impact.

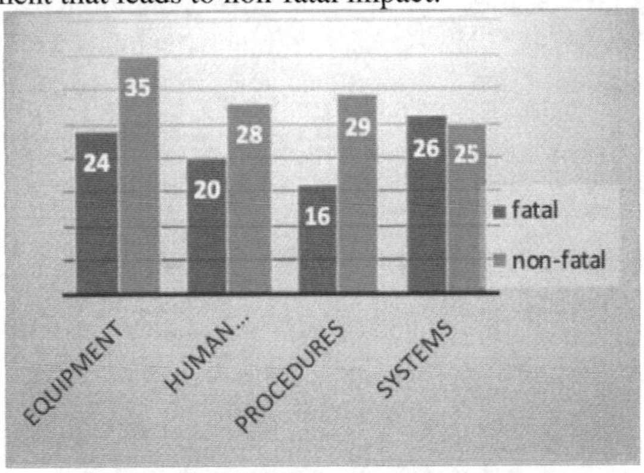

Figure 3. PAIRS factors (per impact).

In Figure 4, it displays the top five aviation accidents which are the *Collision, Crash, Engine Failure, Excursion and Gear Failure* that happened on all phases of operation covering *Inflight, Landing and Take-Off.* It reports that the "Engine Failure" is the top cause of aviation accidents that normally happens during the "Landing" phase of operation in any reported aircraft accidents.

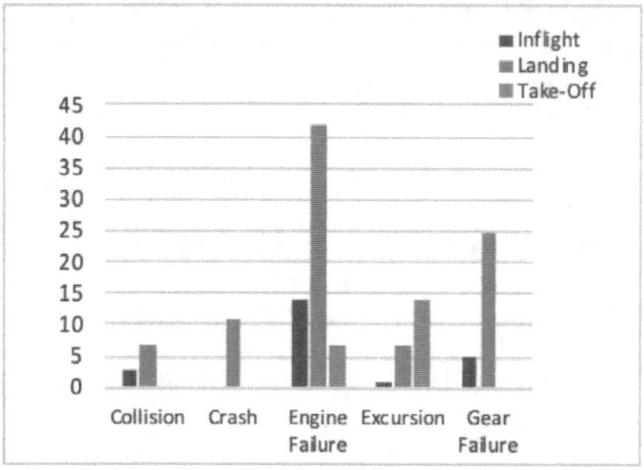

Figure 4. Top five aviation accidents occurrence (per phase operation).

For figure 5, it displays the Aviation Accidents Phase in the whole year from 2008-2017 and covers all phases of operation on which the aviation accident occurred. The diagram shows that the year of 2013 during the Landing phase of operation is the most frequent number of aviation accidents. Compared to the three phases of operation, *Take-Off* is the least frequent phase of operation in all the years.

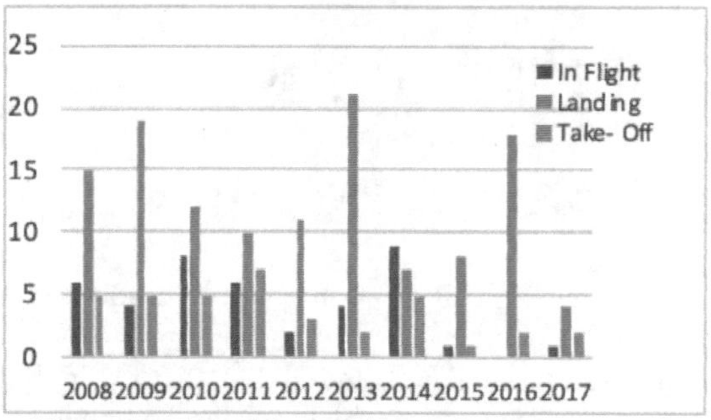

Figure 5. Aviation accidents phase (per year).

For figure 6, it shows the Aviation Accident Weather Conditions can be classified into cloud, visibility and wind. In the pie chart, it shows that the frequent number of aviation accident happens in the windy condition that has 37% and the least frequent accident happens during a cloudy weather which only consists of 30%.

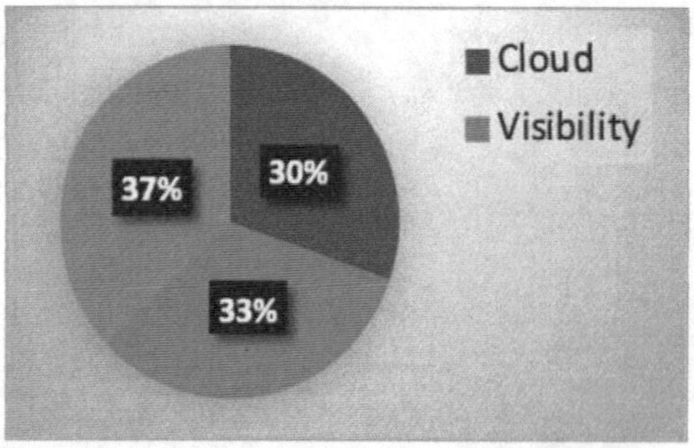

Figure 6. Aviation accident weather conditions.

IOP Conf. Series: Materials Science and Engineering **946** (2020) 012005 doi:10.1088/1757-899X/946/1/012005

In Figure 7 below shows the top five locations where the aviation accidents occurred in between the years of 2008-2017. The top five places are: Aklan, Bulacan, Davao, Manila and Palawan. The dataset includes all the fatal and non-fatal accidents, all the pairs factors, and all-weather conditions. The results show that among the five, the peak of aviation accident happened in the place of Davao and it happened in the year of 2011. It concludes that Davao has consistent increase with the number of accidents every year.

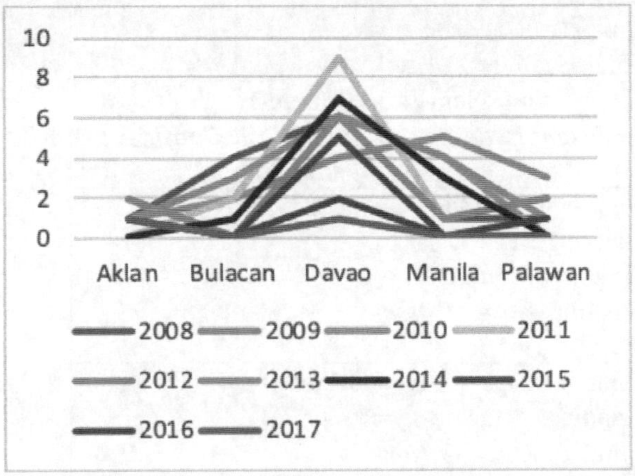

Figure 7. Top 5 locations of aviation accidents.

4.3 Predictive Analytics

The Predictive Analytics uses statistical models and forecasts to answer the question *"What could happen?."* It gives answers that move beyond using historical data as the primary basis for all decisions rather it helps the leaders to anticipate likely scenarios that plan ahead than just reacting to a previous event that occurred. It utilizes models for predicting events. Nevertheless, it does not recommend any actions. This study would be using forecasting and simulation to help enhance the views of the decision makers to make more informed decisions [17]. It will apply the concepts of Association Rule Mining and Time Series Forecasting that would clearly define below.

4.3.1 Association Rules Mining

In this paper, the frequent pattern growth approach (FP growth) is used on which it finds association rules (frequent pattern) and modify it to determine the frequent item sets [4]. The objective of this is to find the patterns in the given aviation accident and which of these association rules are form based on the its minimum support, confidence, and lift. It then helps address the objective of the paper which is to identify associated factors of frequent incidents to the goal of accident prevention. These may assists the value of the reports produced by *PAIRS* as their incident reporting system and can help contribute insights that can be part of the State Safety Program (SSP) and Safety Management System (SMS) of every aviation organization certified by CAAP [3].

The attributes used in the association rule mining are the *factors, year, type_of_occurance, phase_of_operation, weather_condition, impact and province_of_occurance.* All the instances are in a nominal type. To be able to subject the instances to association rule mining, the values required is binomial. Within the preprocessing these data, all the nominal values are converted into numerical version then binomial using the operator commends that is available in the datamining tool, RapidMiner.

The RapidMiner was the tool used for finding the frequent patterns. In validation of the patterns created, there is the minimum support, confidence and lifts measures that was used. The *support* describes how frequent these patterns occur in the dataset. For the *confidence,* it illustrates about how strong those associations are while the *lift,* is used to test the interestingness in determining if the values of the antecedent have a positive effect on the consequent. Thus, the lift value that is greater than 1 will result to a more positive correlation.

IOP Conf. Series: Materials Science and Engineering **946** (2020) 012005 doi:10.1088/1757-899X/946/1/012005

Table 4. FP-Growth results for aviation accidents frequent patterns.

Conclusion: Fatal (impact)			
Frequent Patterns	***min support***	***Confidence***	***lift***
Phase_of_operation = Landing, type_of_occurance = Engine failure → Fatal	0.113	0.548	1.293
Conclusion: Non-Fatal (impact)			
Frequent Patterns	***min support***	***Confidence***	***lift***
Phase_of_operation = Landing, weather_condition= Visibility → *Non-Fatal*	0.128	0.591	1.025
Phase_of_operation = Landing, factors = human factors → *Non-Fatal*	0.084	0.607	1.053
Phase_of_operation = Landing, weather_condition = Cloud factors → *Non-Fatal*	0.118	0.632	1.096
Phase_of_operation = Landing, type_of_occurance = Gear Failure → *Non-Fatal*	0.084	1.180	1.096
Phase_of_operation = Landing, factors = procedures Failure → *Non-Fatal*	0.103	0.700	1.215
Factors = equipment, province_of_occurance = Davao → *Non-Fatal*	0.069	0.700	1.215
Weather_condition = Visibility, factors = equipment → *Non-Fatal*	0.074	0.714	1.239

In this approach, Table 4 shows the FP-Growth was used to discover the best association rules with interesting aviation accidents attributes. The threshold for min support is set to 0.15 and confidence level is set to 0.5 accordingly. This results to have 8 frequent patterns with the impact as the consequent. As seen in the table, there is only one frequent pattern for fatal impact this can imply that this type of impact is not normally happening. It can be seen that non-fatal incidents produces 7 frequent patterns that associate in different attributes from the dataset. Most of the patterns shows that the aviation accident often occurs during the Landing phase of aviation operation and it is matched with different *PAIRS* factors. Hence, for the non-fatal impacts the greater positive correlation based on lift has the value of 1.239 resulted from the combination of Visibility and equipment as factors that leads to non-fatal accidents.

4.3.2 Time Series Forecasting

For this paper, time series forecasting was used with the help of Weka datamining tool. The results was extracted and compared to different algorithm namely: Linear Regression (LR), Gaussian Processes (GP), MultiLayer Perceptron (MP), and Sequential Minimum Optimization Regression (SMOreg) to predict the future trends for the aviation accident for one year in the Philippines. To measure its accuracy, it used the concept of Mean Absolute Error (MAE). MAE is a forecasting accuracy measure to help choose the best-suited method for the aviation accident dataset extracted. The methods used only to look for one year forecast.

The experimental results shows that one year periodicity generated the minimum MAE and it shows the percentage error in each method of the forecasted year which is seen in table 6. By comparing the results, the SMOReg offers the ability to predict the number of aviation accidents in most PAIR factors more accurately compared to other techniques (see table 5). The PAIR factors that got the least MEA are systems and procedures. The forecast value using the SMOreg method is seen in the table 6. The predicted aviation accidents range is 8-30 and the computed MEA for prediction is 0.88 which is low. Hence, it implies that the forecasting method generated better performance. It can be used to reduce the error percentage in predicting the future aviation accidents and it increases the chances of the aircraft officers to predict more accurately.

Table 5. Predicted PAIR factors values (one year).

PAIR factors	LR	GP	MP	SMOReg
Equipment	30.19	25.65	24.59	27.61
Human factors	6.44	17.11	13.07	12.09
Procedures	24.6	24.74	22.87	31.33
Systems	7.24	13.5	30.79	8.16

Table 6. Percentage of error.

Percentage of Error	LR	GP	MP	SMOReg
Correlation Coefficient	1	0.83	0.98	0.99
Mean absolute error	3	17.5	4.06	0.88
Root Mean squared error	5	21.68	5.09	1.55
Relative absolute error	0	55.84	12.95	2.8
Root relative squared error	0	57.01	13.4	4.08

4.4 Prescriptive Analytics

The Prescriptive Analytics uses optimization and simulation to answer the question *"What should we do?."* It is the next method based on the Analytics Value Chain Operational Framework that suggests actions based on the descriptive and predictive analysis of complex data. It results to development of business rules, organization models, comparisons and optimizations based on the outcome of the Predictive Analytics. This study will help prevent occurrence of aviation accidents in the country [17].

In Association Rule Mining, the frequent pattern growth approach (FP growth) algorithm finds the common occurring aviation accident patterns. The results showed that non-fatal impact aviation accidents has a high probability to be from equipment and procedures failures that is associated with frequent engine failures (Landing – Gear Failure – Equipment). The implications of the high rate of failures during the landing phase of operation together with high rate of failures in equipment factor (engine and gear failure) to avoid aviation accidents with fatal impact indicate that solving this problem alone could remedy a majority of aviation accident both for non-fatal and fatal accident impacts.

Another result showed that aviation accidents with non-fatal impacts are more likely to happen in the future during the aircrafts landing phase of operation and the weather condition remains visible and not raining. This is the second most frequent occurring pattern of aviation accident in the dataset is in landing phase with weather circumstances (Landing- Weather - Human Factors). This implies that the impact of accident was not fatal for a visible weather and human factors is the only element that can lead to aviation accident. The importance of understanding the implications of weather and formulating educated in-flight decisions regarding weather must be understood in different levels. In several instances, the pilots without training found themselves in circumstances of deteriorating weather. Instead of choosing alternate pilot, pilots flew into the weather, often lost situational awareness, became spatially disoriented or attempted to land while on unstable approach. In certain circumstances, attempting to land from unstable approach could be considered intentional non-compliance. However, nearly all of the weather-related cases, poor decision-making was demonstrated [18].

For the application of time series forecasting, the SMO regression was used to forecast the number of aviation accidents in all PAIRS factors more accurately compared to the other techniques such as Gaussian processes, linear regression, and Multilayer Perceptron. The outcome of these facilitates the prevention and mitigation of aviation accidents in the future time. Thus, with the help of this study it can help decrease the number of aviation accidents in the Philippines and increase the effectivity and efficiency of aviation accident data analysis.

5. Conclusion

Aviation community in the Philippines still faces challenge on how to maximize the data produced by PAIRS both for incidents and accidents to help prevent aviation accidents. With the help of Open Data, it makes the aviation incidents and accidents scenarios be opened to the public for different span of years. This research study proves that there are associated factors of frequent incidents to the goal of accident prevention. It is important to consider all the associated factors because it has the vital information on the cause of aviation accidents and to help provide the impacts of the said accidents. It uses the Association Rule mining with the help of frequent pattern (FP-growth) algorithm to find the frequent aviation accidents. For the time series forecasting methods, the SMO regression was found to forecast the number of accidents on each PAIR factors accurately. The result created important points in improving the aviation operation and assists in better decision-making to help create policy and future improvement for aviation-safety in the country.

Acknowledgments

Author P.P.R. Abesamis would like to express her gratitude to Dr. Remedios de Dios Bulos, co-author and professor under the course Information Systems Planning, Design, Analysis and Databases, for the opportunity and data analytics knowledge that has been useful in the research paper; to Mr. JP Acuna for recommending the challenge areas that government faces in handling big data; to co-author Ms. Michelle Ching for guiding me with Data Analytics and research principles; to Mr. Neil Datu and Ms. Michelle Cortez, for sharing their knowledge and insights on Data Analytics together with the tools of Weka and RapidMiner and motivating me to finish it; Thank you De La Salle University Manila that provided financial assistance to present this in a conference and publish it in a journal and last but not the least, to our Almighty Father for providing us strength, wisdom, knowledge and motivation to finish this meaningful research study.

References

[1] J. S. S. C. K. Z. Clinton V. Oster, "Analyzing aviation safety: Problems, challenges, opportunities," Research in Transportation Economics ELSEVIER, pp. 148-164, 2013.

[2] J. T. Luxhoj, "Predictive Safety Analytics for Complex Aerospace Systems," Procedia Computer Science (Elsevier), pp. 331-336, 2013.

IOP Conf. Series: Materials Science and Engineering **946** (2020) 012005 doi:10.1088/1757-899X/946/1/012005

[3] CAAP, "Republic of the Philippines Department of Transportation (CIVIL AVIATION AUTHORITY OF THE PHILIPPINES)," Civil Aviation Authority of the Philippines, [Online]. Available: http://www.caap.gov.ph/?page_id=2180. [Accessed 10 December 2018].

[4] P. N. G. P. A. S. Rakesh Kumar Soni, "An FP-Growth Approach to Mining Association Rules," International Journal of Computer Science and Mobile Computing (IJCSMC), vol. 2, no. 2, pp. 1-5, 2013.

[5] S. Y. J. L. a. M. Z. Yi Zeng, "Research of Improved FP-Growth Algorithm in Association Rules Mining," Scientific Programming, vol. 2015, 2015.

[6] P. Tiwari, Accident Analysis by Using Data Mining Techniques, 2017.

[7] E. B. a. P. O. Zohreh Nazeri, "Experiences in Mining Aviation Safety Data," in ACM SIGMOD, 2001.

[8] D. Fengyi and L. Zhenyu, "An ameliorating FP-growth algorithm based on patterns-matrix," Journal of Xiamen University (Natural Science), vol. 44, pp. 629-633, 2005.

[9] Y. Yang and Y. Luo, "Improved algorithm based on FP-Growth," Computer Engineering and Design, pp. 1506-1509, 2010.

[10] S. Kumar, "Analyzing Road Accident Data Using Association Rule Mining," IEEE, 2015.

[11] D. Z. a. K. Jianga, "Application of Data Mining Techniques in the Analysis of Fire Incidents," International Symposium on Safety Science and Engineering in China, 2012.

[12] Balogun.O.S., "International Journal of Advanced Research," USE OF TIME SERIES ANALYSIS OF ROAD ACCIDENT DATA IN LAGOS STATE, vol. 2, no. 5, pp. 1045-1059, 2014.

[13] D. A. Arora, "Data Mining Techniques and Tools for Knowledge Discovery in Agricultural Datasets," IASRI, 2011.

[14] J. H. a. M. Kamber, Data Mining: Concepts and Techniques, Morgan Kaufmann Publishers, 2000.

[15] Informatica, "What is Data Analytics?" Informatica, 2018. [Online]. Available: https://www.informatica.com/services-and-training/glossary-of-terms/data-analytics-definition.html#fbid=hYXDmWmJvom. [Accessed 12 December 2018].

[16] "Data Analytics," technopedia, 2018. [Online]. Available: https://www.techopedia.com/definition/26418/data-analytics. [Accessed 30 November 2018].

[17] IBM, "IBM Software Thought Leadership White Paper," Descriptive, predictive, prescriptive: Transforming asset and facilities management with analytics, 2013.

[18] B. N. Branham, "Analysis of Fatal General Aviation Accidents Occurring from Loss of Control on Approach and Landing," EMBRYRIDDLE Aeronautical University Scholarly Commons, August 2013.

IOP Conf. Series: Materials Science and Engineering **946** (2020) 012006 doi:10.1088/1757-899X/946/1/012006

Multipoint intermittent fault detection of electronic interconnections using channel sounding techniques

Husain Kanchwala[1] **and Syed Wakil Ahmad**[2]

[1] Warwick Manufacturing Group, Univeristy of Warwick, Coventry CV4 7AL, United Kingdom
[2] Department of Mechanical Engineering Sciences, University of Surrey, Guildford GU2 7XH, United Kingdom

E-mail: husain.kanchwala@warwick.ac.uk

Abstract. Intermittent faults (IFs) in electrical interconnection are short duration transients. These faults are random, non-predictable and non-stationary in nature making them hard to troubleshoot using traditional equipment. They can be detected only if the test equipment has test coverage both in time and area at that instant. This paper proposes a novel approach of intermittent fault detection by injecting a wideband signal and using channel-sounding techniques. A novel Direct Sequence Spread Spectrum (DS-SS) technique is presented for multipoint IF detection and health monitoring. Digital communication sounding technology and its methods devised for detection and classification of an IF, and channel characterisation by its transfer function is discussed in detail. A combination of transmitter and receiver provide an effective tool for continuously monitoring of the multi-point interconnect. The proposed method has been tested using a functional prototype having pre-induced IFs. Intermittent signal is generated by applying external vibration on the connector and intermittency is detected by acquiring and processing the propagated signal. The results demonstrate the effectiveness of the method in detection and classification of intermittent interconnections.

Keywords: fault diagnosis, maintenance engineering, condition monitoring, integrated circuit interconnections, circuit faults, vibrations, electric connectors.

1. Introduction

With increasing complexity of digital electronic circuits, the probability of occurrence of faults has also increased. The most common fault in these circuits occur in electronic interconnections. Faults cause the system to operate far away from its optimal point and reduces its efficiency. It is critical to identify them due to saftey concerns specially in aerospace applications [1].

System faults are broadly classified into three categories: permanent, transient and intermittent faults (IFs) [2]. IFs are electrical spikes developing due to ageing interconnects, cuts, rubs, or loose contacts. They occur in an unpredictable manner and are safety critical [3].

Manufacturing imperfections, poor design and system degradation are main causes of IFs which are in turn precursors of permanent failure [4, 5]. These are random and non-reproducible incidents, and are not usually identified by visual or traditional instruments for electronic/electrical interconnects. When the fault is intermittent, it could not be present at the time of testing and will be tagged as No Fault Found (NFF) [6]. Industries are suffering from NFF failures due to lack of necessary levels of confidence and efficiency of the methods used in detecting these events. Traditional test equipments used to detected anomalies like multi-meters

Content from this work may be used under the terms of the Creative Commons Attribution 3.0 licence. Any further distribution of this work must maintain attribution to the author(s) and the title of the work, journal citation and DOI.
Published under licence by IOP Publishing Ltd

IOP Conf. Series: Materials Science and Engineering **946** (2020) 012006 doi:10.1088/1757-899X/946/1/012006

detect only steady or invariant signals. While digital oscilloscopes can detect changing signals based on the sampling function, problem with an IF is that it occurs for a very small duration, making it difficult to detect unless a very high sample rate it used which is beyond the capabilities of traditional digital oscilloscopes. A 2012 survey of 80 aerospace organizations ranked IFs as the highest perceived cause of NFF [7]. The survey results gives a strong motivation to develop a novel algorithm for electronic interconnection faults using channel sounding techniques.

2. Methodology

The current state-of the-art IF detection during maintenance testing includes latching continuity testing, analogue neural network technology and time domain reflectometry. Reflectometry is widely used for high power electrical wirings and could not be used for interfaces, lose solder joints and other electronics circuits. It relies on transmitting electromagnetic waves across the wire and in turn observing the reflections R.

$$R = \frac{(Z_1 - Z_2)}{(Z_1 + Z_2)}, \tag{1}$$

where Z_1 and Z_2 are impedances of two electrical mediums. These reflections depend on the variation of impedance in the wire system. Time duration between the incident and reflected wave is used to locate the fault. Magnitude of reflections is used to determine if it is a potential fault or not. The drawback of reflection based techniques is that any change in the wire material (e.g., connection in circuit) reflects the incident waves resulting in erroneous fault determination. Moreover, they requires high voltage pulses which can't be applied to low voltage digital circuits.

Recently, direct-sequence spread-spectrum (DS-SS) modulated signals are used instead of high voltage signals in employing digital signal processing techniques to locate faults [8]. It uses low voltage signals and does not interfere with online signals and could be used in-situ, but it still suffers from reflection occurring at all points of interconnections in the circuit making it unsuitable for electronic circuits like PCBs, solder joints, interfaces, and similar interconnections.

Kim *et al.* used DS-SS frequency shift keying (FSK) modulation technique to overcome reflectometry issue of reflections and false alarms [9]. Although the false alarms problem was solved but still it could not be used for mesh and complicated architecture of multipoint and does not give channel characterisation or degradation information. Taylor and Faulkner initially proposed DS-SS modulation technique, and outlined optimal signal processing and frequency domain correlation for the on-line test in high voltage power line carrier [10]. Furse *et al.* employed a slightly different use of the spread spectrum for detecting live wire problems [11]. This paper presents a novel method of IF detection and characterisation to overcome the above mentioned issues. The DC signal is spread with DS-SS and send into interconnection system for IF detection and channel measurement. This could be used for multipoint interconnections and to diagnose wire health after de-spreading to retrieve an intermittent signal at output of the system that are not extractable using traditional equipments like oscilloscope, reflectometry,neural network based bespoke equipment or voltmeter. The transients caused by IFs in the wire would disrupt the DS-SS signal sent over interconnection from the transmitter. The DS-SS signal retrieved at the receiver end would contain intermittent signal information. IF will eventually be deducted from the signal by de-spreading i.e. multiplying it with the PN-code. Only same PN-code will de-spread the signal which has been used for spreading, each node will have unique PN-code. The communication technology and its devised method for detection and extraction of faults information in terms of duration and occurrence frequency, and channel impulse response for characterisation of wire channels is described in detail. The simulation of the prototype system is performed and experiments were conducted on the rig. Finally conclusions and future scope of the work has been discussed.

IOP Conf. Series: Materials Science and Engineering **946** (2020) 012006 doi:10.1088/1757-899X/946/1/012006

3. Communication Sounding Approach for Intermittent Fault Detection

System impulse response technique is used to get the frequency response of the system. This is used in many applications and most electronics system are designed with the consideration of transfer function or impulse response. This is also used to get channel response to see behaviour of channel over frequencies of interests for reliable communication. This concept could be used to get frequency response of wire or other interconnection system and see behaviour of response over wider band of frequencies to estimate deterioration or degradation The impulse response will give the channel response over wide range of frequencies. Any crack or degradation could be detected by impulse response. Small wire degradation might not effect DC or low frequencies but higher bandwidth response can be used to estimate wiring and interconnection system. In time domain, the channel can be described as complex impulse response $h(t, \tau)$ as shown in Figure 1.

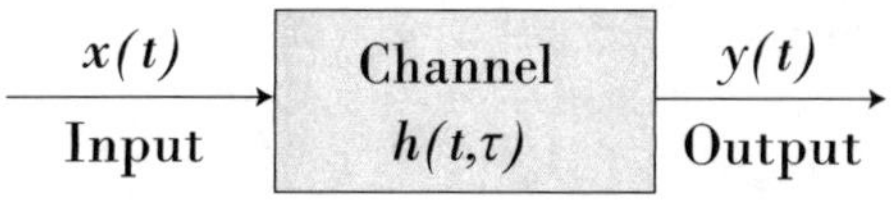

Figure 1. Channel impulse response.

$$y(t) = h(t, \tau) * x(t). \tag{2}$$

Equation (2) shows the relation between input and output of channel, where, $x(t)$, $y(t)$ are input/output, and impulse response $h(t, \tau)$ and '$*$' is the convolution. To de-convolve $x(t)$ from $y(t)$ in time domain is bit complicated but This could be expressed in the frequency domain as

$$Y(f) = H(f) \cdot X(f) \implies H(f) = \frac{Y(f)}{X(f)}. \tag{3}$$

$X(f)$, $Y(f)$ are the Fourier transforms of $x(t)$ and $y(t)$. It is more convenient to calculate impulse response or the transfer function for discrete time signal using $z-$transforms defined as

$$X(z) = \sum_{n=0}^{\infty} x(n)z^{-n}; \quad H(z) = \frac{Y(z)}{X(z)} = \frac{y_0 + y_1 z^{-1} + \ldots + y_{n-1}z^{-(n-1)} + y_n z^{-n}}{x_0 + x_1 z^{-1} + \ldots + x_{n-1}z^{-(n-1)} + x_n z^{-n}}. \tag{4}$$

Where $X(z)$ is $z-$transform of $x(n)$, and $H(z)$ is the transfer function.

4. Orthogonal codes

A pseudo noise code (PN code) has a similar spectrum as that of a random noise but it is generated deterministically. PN codes are generated by using shift registers with modulo-2 arithmetic. Most widely used sequences in DS-SS systems are Gold codes, Kasami codes, Barker codes, and Hadamard codes [12]. Hadamard codes are perfectly orthogonal and are optimal codes to avoid interference among users in the link from base station to terminals. This paper uses a Hadamard code for channel sounding. To obtain the orthogonal set, the approach is to select these codes from rows or columns of a Hadamard matrix. A Hadamard matrix is $2^k \times 2^k$ square matrix of elements belong to $(1, -1)$ set. The simplest matrix of 2 orthogonal Walsh-Hadamard sequences is given by

$$H_{(2)} = \begin{bmatrix} 1 & 1 \\ 1 & -1 \end{bmatrix} \tag{5}$$

The code of user 1 is the first column and the code of user 2 is the second column. Clearly the first column $(1, 1)$ is orthogonal to the second column $(1, -1)$. This matrix is eventually extended using a recursive technique. For 2^N users, the matrix is found from the code matrix for $2^{(N-1)}$ users, according to

$$H_{(2N)} = \begin{bmatrix} H_N & H_N \\ H_N & -H_N \end{bmatrix} \tag{6}$$

Orthogonal codes are used for spreading in communication systems where the receiver is perfectly synchronized with the transmitter.The de-spreading operation is ideal, as the codes are completely de-correlated. Hadamard matrices are square matrices with entries either +1 or −1. If N is a nonnegative power of 2, the $N \times N$ Hadamard matrix H_N, is defined recursively as given in Equation (6). H_N has the property that

$$H_N H_N^T = N I_N, \tag{7}$$

where I_N is the identity matrix of size $N \times N$. The code generator block outputs a row of H_N. The output is bipolar. The code length parameter N must be a power of 2 and is specified by the user. Moreover the index of the row of the Hadamard matrix, which is an integer in the range $[0,\ 1, \ldots,\ N-1]$, called as the code index parameter is also user specified.

5. DS-SS Sounding technique for IF Detection

DS-SS sounding technique for IF detection in multi-points electronic interconnection system sends a DC constant unit signal input and spreads it with pseudo-random noise (PN), orthogonal codes to wider band. Hadamard orthogonal sequence is used for spreading in this system because receiver is perfectly synchronized with the transmitter. De-spreading operation is ideal, as the codes are completely de-correlated and PN is completely removed. As input to diagnolize the system is a constant DC signal, the output will be a transfer function that gives channel impulse response using channel sounding technique. The complete system is shown schematically in Figure 2. A carrier signal based approach to detect IF has been investigated earlier by one of the authors in [13]. This approach is an advanced version of similar underlying ideas.

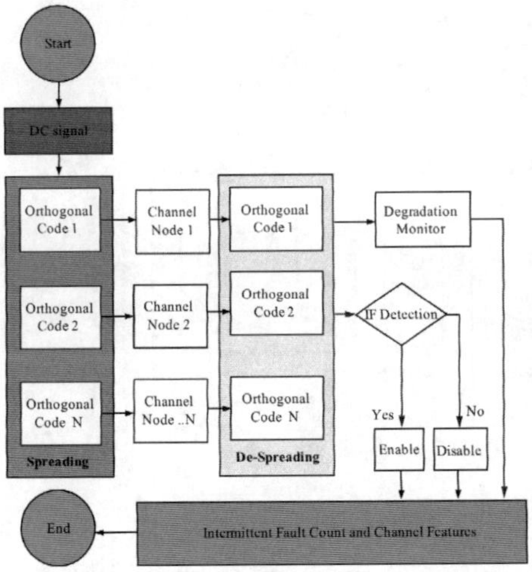

Figure 2. Block Diagram of Multi Node IF Detection using Channel Sounding.

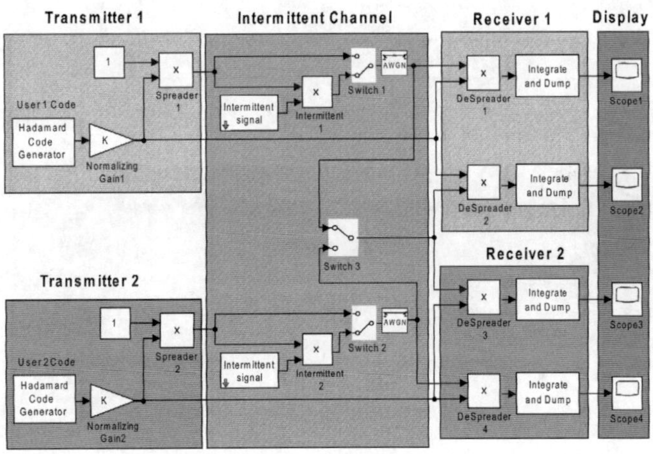

Figure 3. Two point intermittent fault detection model.

6. Simulation

The simulation model of the DS-SS sounding technique is developed in MATLAB/Simulink (see Figure 3). It could be modeled with as many test points as required but only two test points has been simulated to show how it could be used to test multi point interconnections. It consists of two transmitters, two receivers, and two test nodes of IF. In our model we have also included Additive White Gaussian Noise (AWGN) to make it more realistic. We have simulated it with two intermittent connection and showed how DS-SS detects IF and it also shows that cross talk with other point does not de-spread, thus eliminating the false alarm in multi node configuration.

IOP Conf. Series: Materials Science and Engineering **946** (2020) 012006 doi:10.1088/1757-899X/946/1/012006

The system consists of three sub blocks transmitter, channel and receiver. Each transmitter and receiver used the same orthogonal codes to spread/de-spread input signal as described in previous section. Every transmitter spreads a constant voltage signal with different Hadamard codes and propagates through an electronics interconnection system for IF and feature detection. De-spreading unit removes pseudo-noise (PN) by multiplying it with same generated PN code.

7. Simulation results

To monitor the signals at inputs/outputs, scopes are connected to the input and output as shown in Figure 3. Scope 1 and 2 are connected with node 1 with same PN codes,3 is connected to output with different PN code's signal than transmitter while 4 is connected to perfectly synchronised PN code with transmitter. A constant value input signal of unit magnitude is connected to all inputs. Intermittent signals from node 1 are regenerated by removing DS-SS signal are shown in Figure 4. As Scope 3 is connected to different PN code it is not able get intermittent signal and the output is noisy. Scope 4 shows the perfectly synchronised PN code.

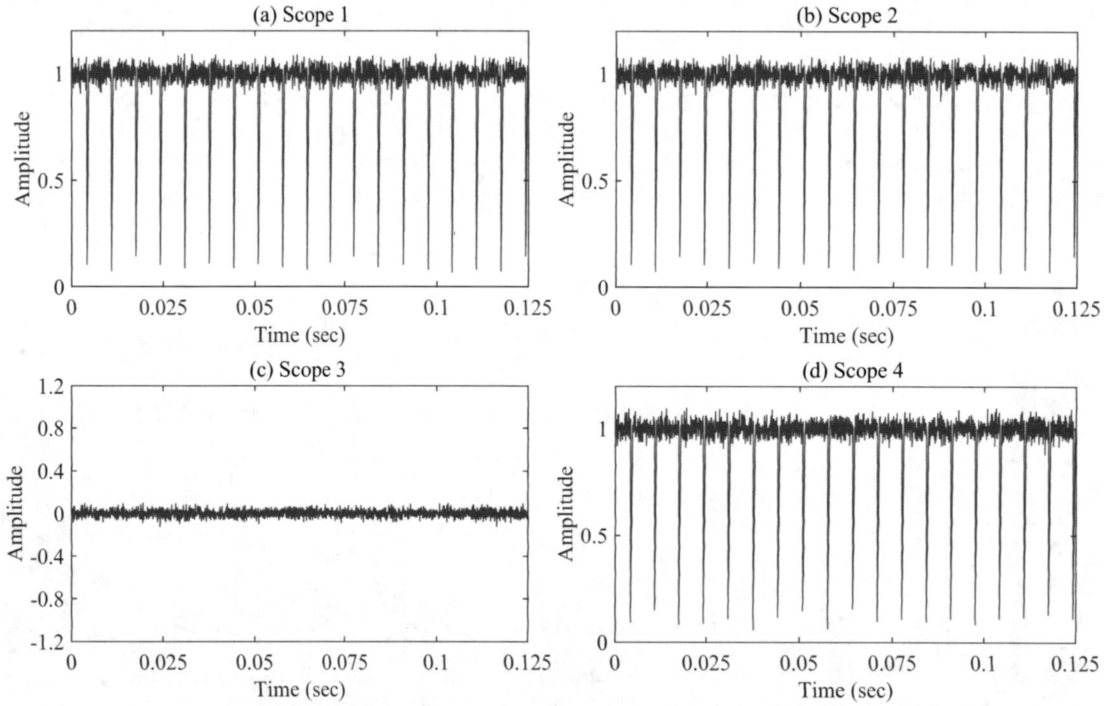

Figure 4. Simulation results: (a) Scope 1: Signal showing channel IF. (b) Scope 2: Channel IF. (c) Scope 3: Poor de-spreading due to wrong orthogonal code. (d) Scope 4: Perfectly synchronised PN code.

8. Validation

This algorithm has been validated by acquiring data to Simulink model from National Instrument (NI) 6363 X-series data acquisition system. This system has both analogue and digital input/outputs. Experimental setup is shown in Figure 5. DS-SS signals are generated in Simulink model and sent to interconnections system through NI-6363. Ethernet male and female socket on shaker has been used as test rig, to introduce intermittency in DS-SS signals. The signals are acquired back by same NI system for de-spreading and intermittent feature extraction. For further details regarding the experimental setup see [14]. This intermittent fault detection system has been described by using a block diagram as shown in Figure 6. Two PN

IOP Conf. Series: Materials Science and Engineering **946** (2020) 012006 doi:10.1088/1757-899X/946/1/012006

codes are used for two node IF detection. A DC unit signal is spread and DS-SS signals are sent to a multipoint ethernet socket on the test rig. PN codes are removed by multiplying/de-spreading to get intermittent interconnection information as described in simulation results.

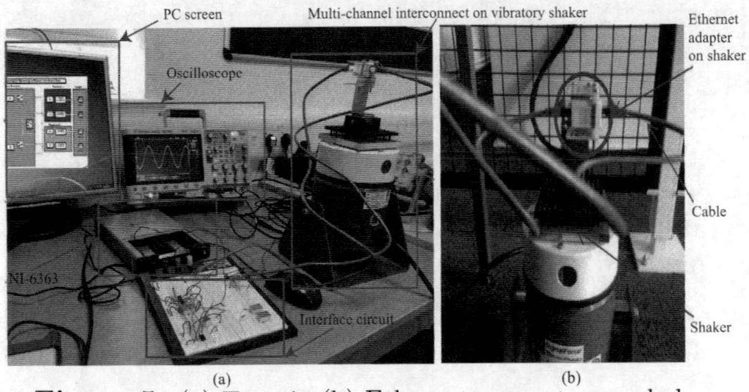

Figure 5. (a) Test-rig (b) Ethernet connector on shaker.

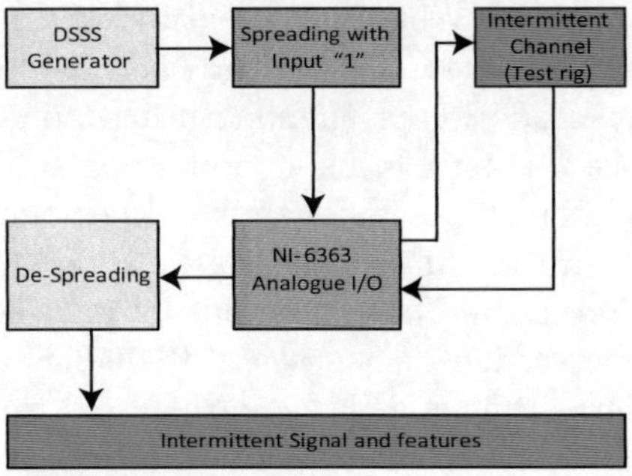

Figure 6. Block diagram of testrig.

Same Simulink model is used for this experiment as described in the Simulation section earlier. The results are described in Figure 7. Scope 1 and 2 shows the results of same interconnection loop at node 1 with same PN codes synchronisation, whereas Scope 3 and 4 shows the results of second node but with two different PN codes.

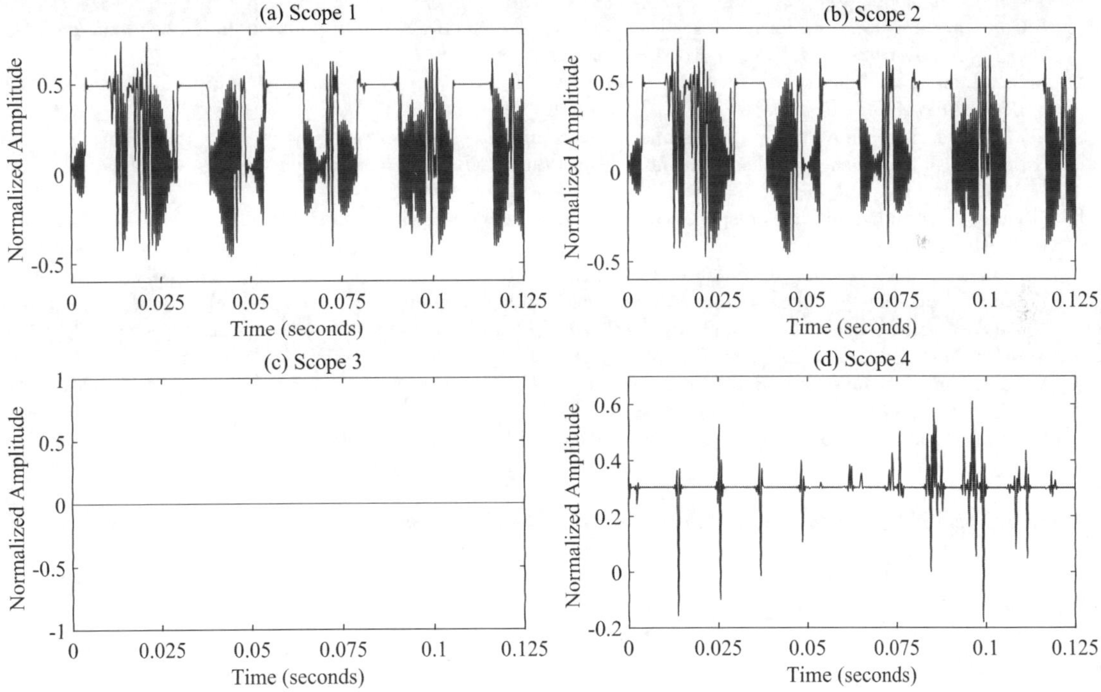

Figure 7. Test results: (a) Scope 1: Node 1 with same PN code de-spreading. (b) Scope 2: Node 1 with same PN code de-spreading. (c) Scope 3: Node 2 with different PN code de-spreading. (d) Scope 4: Node 2 with same PN code de-spreading.

IOP Conf. Series: Materials Science and Engineering **946** (2020) 012006 doi:10.1088/1757-899X/946/1/012006

9. Conclusions

Multipoint intermittent interconnections can be detected by using DS-SS schemes and it could be classified with respect to band of intermittence that linked with external vibration. It also reduces human factor's mistake to put wrong node for IF detection because only same PN code could only be used to extract IF information. The characteristics of IF could also be detected with respect to amplitude and frequency and these could be used to predict the interconnection degradation or manufacturing faults. It is low power and very efficient technique that could be used in-situ for multi-nodes. This method is generalized and filter banks can be used to segment different possible bands of spectrum. In future we would focus on adding wavelets banks to segment different possible bands of spectrum for further analysis of intermittent signal. This work could be carried out by wavelets transforms, or Short Time Fourier Transform (STFT); to determine the sinusoidal frequency and phase content of local sections of a signal as it changes over time and this could be used to detect intermittency, and to find root causes.

References

[1] A Correcher, E Garcia, F Morant, E Quiles, and R Blasco-Gimenez. Intermittent failure diagnosis in industrial processes. In *2003 IEEE International Symposium on Industrial Electronics (Cat. No. 03TH8692)*, volume 2, pages 723–728. IEEE, 2003.

[2] J Guilhemsang, O Héron, N Ventroux, O Goncalves, and A Giulieri. Impact of the application activity on intermittent faults in embedded systems. In *29th VLSI Test Symposium*, pages 191–196. IEEE, 2011.

[3] Q Li, K Lyu, J Qiu, and G Liu. Research on intermittent failure re-presentation of electrical connector based on accelerated test. *Proceedings of the Institution of Mechanical Engineers, Part O: Journal of Risk and Reliability*, 233(3):317–327, 2019.

[4] A Correcher, E García, F Morant, E Quiles, and L Rodríguez. Intermittent failure dynamics characterization. *IEEE transactions on reliability*, 61(3):649–658, 2012.

[5] B Steadman, F Berghout, N Olsen, and B Sorensen. Intermittent fault detection and isolation system. In *2008 IEEE Autotestcon*, pages 37–40. IEEE, 2008.

[6] SW Ahmad, S Perinpanayagam, M Samie, and I Jennions. A novel intermittent fault detection algorithm and health monitoring for electronic interconnections. *IEEE Transactions on Components, Packaging and Manufacturing Technology*, 6(3):400–406, 2016.

[7] G Huby. No Fault Found: Aerospace Survey Results. *Copernicus Technology Ltd*, 2012.

[8] P Smith, C Furse, and J Gunther. Analysis of spread spectrum time domain reflectometry for wire fault location. *IEEE sensors journal*, 5(6):1469–1478, 2005.

[9] C Kim. Detection and location of intermittent faults by monitoring carrier signal channel behavior of electrical interconnection system. In *2009 IEEE Electric Ship Technologies Symposium*, pages 449–455. IEEE, 2009.

[10] V Taylor and M Faulkner. Line monitoring and fault location using spread spectrum on power line carrier. *IEE Proceedings-Generation, Transmission and Distribution*, 143(5):427–434, 1996.

[11] C Furse, P Smith, C Lo, YC Chung, P Pendayala, and K Nagoti. Spread spectrum sensors for critical fault location on live wire networks. *Structural Control and Health Monitoring*, 12(3-4):257–267, 2005.

[12] GPS government Technical report. Pseudorandom Noise (PRN) Expansion.

[13] SW Ahmad, S Perinpanayagam, IK Jennions, and M Samie. A carrier signal approach for intermittent fault detection and health monitoring for electronics interconnections system. 2015.

[14] SW Ahmad, S Perinpanayagam, I Jennions, and S Khan. Study on intermittent faults and electrical continuity. *Procedia CIRP*, 22:71–75, 2014.

An Analytical Approach of Integrating Automated Load Scheduling to a Smart Energy Meter using Differential Evolution Algorithm

Aditya Vikram, Dhairya Karna, Astitva Kumar and Mohammad Rizwan

Department of Electrical Engineering, Delhi Technological University (DTU), Delhi, India

adityavikarm_bt2k16@dtu.ac.in

Abstract. Leading-edge power systems which are employed in present-day scenario are significantly complex, extensively distributed and populous. Rising costs of generating power, ever-increasing demand for electricity and shortage of resources have led to introduction of applications of optimal power systems. Hence Smart Energy Meters (SEMs) and highly efficient algorithms are required to facilitate Automated Load Scheduling (ALS). The latest optimization methods for ALS which works on the principle of genetics and natural selection has been presented in the paper. Differential Evolution (DE) is a simple numerical optimization approach that is easy to implement and significantly faster and robust. This article presents a novel mathematical modelling approach for the minimization of the energy consumption. The paper proposes an in depth modelling approach for ALS considering numerous aspects of the DE: initialization, mutation, crossover and selection. Case-studies have been made which include multiple aspects of non-linearities like prohibited zones of operation and transmission losses. The performance of DE based ALS algorithm was implemented using Python and evaluated at the test site in New Delhi which integrates a local renewable energy source (Local PV Grid) with an SEM. The Peak-to-Average (PAR) Ratio was used as a measure of evaluation. The proposed algorithm saved up to 19.42% power in the best-case scenario.

1. Introduction

Smart Energy Meter (SEM) technologies have led to a worldwide change in the aspect of power grid management. The Energy Management (EM) in SEMs is eminently complicated process which needs to be discussed so as to maximise the capability to generate income while minimizing the power consumption of the unit. Over the years, the power grid and SEMs have changed into an advanced system of networks with the purpose of integration of large number of Local Energy Resource (LER) under suitable conditions of the market. [1]. However, one of the major challenges we currently face is the inability of LERs to compete in the present-day market due to their irregularities in energy generation capabilities which puts a limit of their use in grid contribution operations. [2]-[3]. The perfect known solution to this problem is the accumulation of LERs through Smart Grids (SG) and Storage units (Virtual Power Plants (VPP)) providing functionalities like controlling ability, data monitoring and capability to compete in market with the conventional sources of energy. [4].

This paper focuses on the above-mentioned problems of Energy Management so as to provide effective support in the operation of SGs and VPPs. The role of the VPP is the effective control of LERs with the aim to reduce power generation and costs while maximising profits. In a real-time

Content from this work may be used under the terms of the Creative Commons Attribution 3.0 licence. Any further distribution of this work must maintain attribution to the author(s) and the title of the work, journal citation and DOI.
Published under licence by IOP Publishing Ltd

IOP Conf. Series: Materials Science and Engineering **946** (2020) 012007 doi:10.1088/1757-899X/946/1/012007

scenario to successfully achieve this goal the EM should incorporate a massive assortment of resources like different types of Storage Units and Distributed Generation System. [5]. Furthermore, factors like supplier involvement, bids in market, balance restrictions of AC power, along with a proper Demand Response Management (DRM) programs turn the EM problem into a Floating Decimal Non-Linear Problem (FDNLP). [6]-[7]. It is quite known that an FDNLP is complex and hard to solve using classical and basic approaches, but are quite apt to be solved via Evolutionary Algorithm (EA). [8]-[9]. This paper recommends the application of DE to find an optimized solution for the EM problem in the SEM and SG. Differential Evolution provides a simple algorithm for optimization, yet it is effective for global optimization. In the classical form, Differential Evolution is loaded with a set of solutions and it then follows analogous set of computational procedures (like crossover, selection and mutation) after each iteration. The application of the DE was implemented with the help of the scipy library for Python and used for integration at the test site (Energy Utilisation Laboratory, DTU, Delhi, India) between the Local Solar PV grid and the Smart Energy Meter (SEM). The algorithm contains the mathematical model for each operation of the DE Algorithm specifically tailored to the optimization problem. The objective of the minimization of energy consumption along with increment in the amount of energy generation is proposed to be achieved by this use of DE Optimization Algorithm. An exhaustive analytical and successful practical application of DE is provided in this paper and various other applications of DE in various disciplines of science can be found in [10].

2. Differential Evolution

Developed in 1997 by Price and Storn, Differential Evolution (DE), is an optimization technique that is easy to implement as well as quite robust and considerably faster. One of the unique features of operators for evolution in DE entails various components of the present generation adapting to a specified scaled step-size. At the time, numerous variations were introduced that can help differentiate between the process and make the evolutionary process more dynamic e.g. population size of the sample etc. Use of Differential Evolution has been done numerous optimization problems of the power system such as the problem on planning of generation expansion (by Kannan, 2005) and problem of scheduling of hydrothermal scheduling (Subramanian and Lakshminarasimman, 2006). [11]. The focus of this paper is on the implementation of the DE algorithms to the optimization problem of minimization of energy consumption and increment in the energy generation. Differential Evolution determines the unit of optimum operation with the prediction of the demand load in a scheduled time period.

3. Mathematical Model

The DE optimization technique disseminates the total generation requirement among the operation units to minimize the energy consumption and transmission losses thereby increasing the amount of energy generated over a recommended load schedule. The mathematical model determines the five core models for the operation of the DE Algorithm in regards to the optimization problems. The five core models are of objective function for minimization, equation for initialization of the DE Algorithm, mutation, crossover and selection processes.

3.1 Objective Function

The objective function for the problem of minimization of consumption energy can be represented as the minimizing the following function:

$$\text{Optimize } P = \sum_{i=1}^{Ng} PC(i) \tag{1}$$

where Ng is the number of operation units and P is the energy consumption. The power consumption, PC equation is represented in terms of decision variables as follows

$$PC(i) = \sqrt{3}V_{il}I_{il}PF_i \tag{2}$$

where V_{il} is the parameter of line voltage, I_{il} is the parameter of line current and PF_i is the parameter of power factor of the ith unit of operation.

3.2 Initialization

An arbitrary selection of Np individual entities is made on the basis of uniform distribution for the initial population so that all the present variables can encompass the complete search space in a uniform manner. The expression of power generation by the ith PV module is as follows

$$P_i = P_i^{min} + \rho\left(P_i^{max} - P_i^{min}\right) \quad (3)$$

where ρ is a random number of uniform distribution and it belongs to the set of [0,1].

The concept of penalty function is used to tackle to issue of constraint of power balance. This approach decreases the vector fitness in accordance to the parameter of constraint infraction. The corrected expression for objective function to be used for minimization problem is

$$\varphi = \sum_{i=1}^{Ng} (\sqrt{3}\, V_{il} I_{il} PF_i) + \sum_{z=1}^{Nc} \lambda_c |VIOL_z| \quad (4)$$

where λ_c is known as the penalty factor, Nc represents the number of constraints and $VIOL_z$ is the expression for constraint violation.

3.3 Mutation

Generation of new vector in DE is done by summation of the weighted difference between two members to the third member. A new vector is generated along the lines of parent vector in the process of mutation governed by the following model

$$Z_i^{G+1} = Z_p^G + F \times (Z_m^G - Z_n^G) \quad (5)$$

where m and n are random selection and F represents the scaling factor. The value of F ∈ [0,1] and it assures the possibility of quickest convergence and G is the expression for generation number.

3.4 Crossover

The process of crossover produces the upcoming generation from the new vector and the present vector on basis of binomial distribution. The expression of crossover is determined as

$$\hat{Z}_i^{G+1} \begin{cases} Z_{ji}^G & if\ random\ no. > C_R \\ \hat{Z}_{ji}^{G+1} & \end{cases} \quad (6)$$

where C_R is the crossover factor which is user assigned and $C_R \in$ [0,1] and $i = 1,2 \dots N_p$, $j = 1,2 \dots n$.

3.5 Selection

The process of selection is a competition of the parent and the successor one-on-one. If the fitness of the successor is surpassing that of the corresponding parent then the successor replaces the parent. Otherwise, if the fitness of the parent is surpassing that of the corresponding offspring then the retention of the parent takes place.

The expression of choosing between the fitness of the parent and its successor is defined as

$$Z_i^{G+1} = \arg\min\{\varphi(Z_i^G), \varphi(\hat{Z}_i^{G+1})\} \quad (7)$$

$$\hat{Z}_p^{G+1} = \arg\min\{\, \varphi(Z_i^{G+1}), i = 1,2,3 \dots N_p\} \quad (8)$$

where $\arg_{min}$ is the minimum argument and $i = 1,2 \dots N_p$.

4. Results and Discussion

In this paper, the performance of the DE Optimization Algorithm for energy consumption is carried out in Python (scipy library) code on a Core i7 PC while the simulations are carried out on the MATLAB platform. This data is over a scheduled evaluation time period of 15 days from mid-December to starting of January. The related energy consumption, generation and convergence

IOP Conf. Series: Materials Science and Engineering **946** (2020) 012007 doi:10.1088/1757-899X/946/1/012007

characteristics are thereby represented in the Fig 1, Fig 2 and Fig 3 respectively. The laboratory setup is also shown in Fig 4. The SEM are further connected to automated data loggers. The results quite evidently express that the proposed DE Algorithm for the optimization of energy consumption is better than the solution that is expressed in the literature overview. The best DE parameter were observed at F = 1, C = 0.71 and N_p = 38 while the optimized results were procured in 20 seconds.

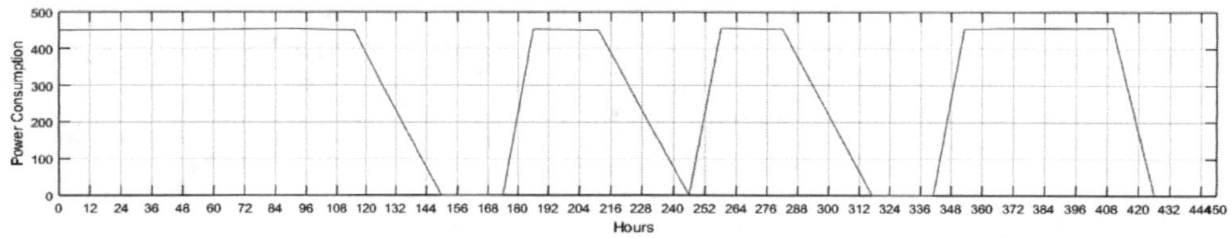

Figure 1. Energy Consumption over scheduled evaluation period of 15 days

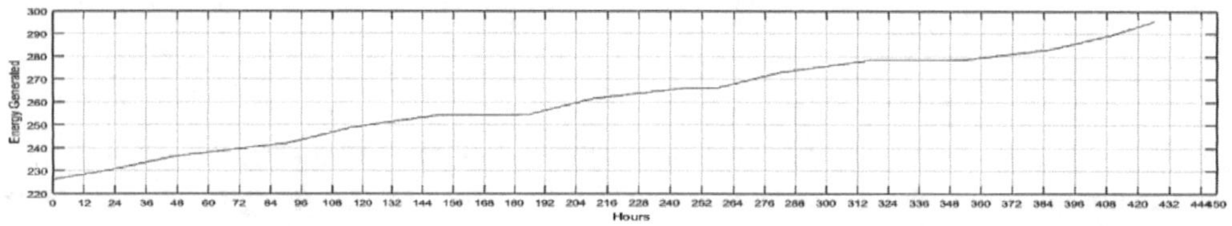

Figure 2. Energy Generation over scheduled evaluation period

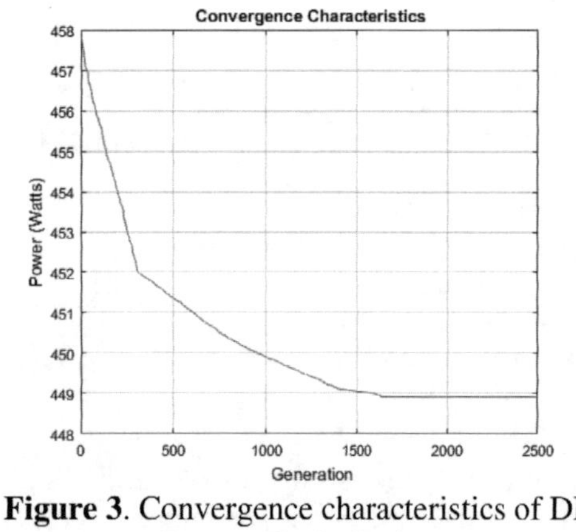

Figure 3. Convergence characteristics of DE Energy Optimization

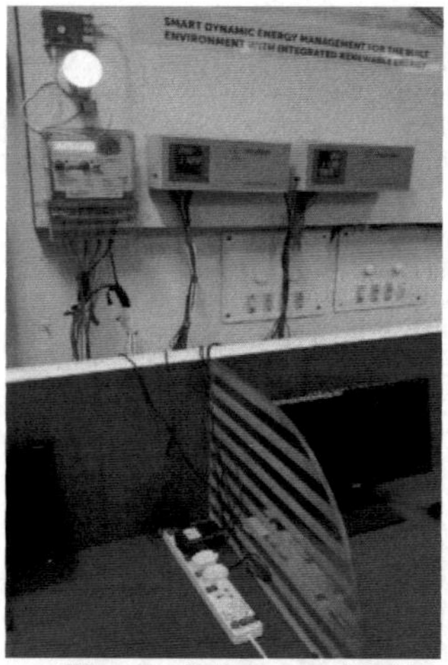

Figure 4. Image of SEM integrated with a PC for DE Optimization

5. Conclusion

In this paper, the integration of Smart Energy Meter and Smart Grids is done with a Differential Evolution Optimization Algorithm for the purpose of optimization of energy consumption and generation via the load present in the at the test site in New Delhi and local renewable energy source (Solar PV) for generation. The results clearly express that the objective of smart energy consumption, reduced transmission losses due to the smart energy consumption and increase in energy generation is successfully achieved. The proposed DE optimization algorithm saved up to 19.42% power in the best-case scenario. The Peak-to-Average (PAR) ratio was used as a parameter for evaluation over the scheduled evaluation period of 15 days. According to the results the best DE parameter was observed at F=1, C = 0.71 and Np = 38 while the optimized results were procured in 20 seconds. This arrangement is financially sound not only for the electricity consumer (in terms of cost and energy saved) but also to the producer (in terms of higher energy generation). Additionally, the reduced value in power consumption from the convergence characteristics and increased generation is an indication to minimized transmission losses which inhibits the hampering of the Transmission and Distribution (T&D) channels. Further implementation in industrial scale controllers and customized software (scipy Python library) is recommended.

References

[1] Ye Yan, Yi Qian, Hamid Sharif and David Tipper. 2013 . A survey on smart grid communication infrastructures: Motivations, requirements and challenges. *IEEE communications surveys & tutorials* 15, 1 (2013), 5-20.

[2] Kumar, A., Rizwan, M. & Nangia, U. A Hybrid Intelligent Approach for Solar Photovoltaic Power Forecasting: Impact of Aerosol Data. Arab J Sci Eng 45, 1715–1732 (2020).

[3] D. Karna, A. Vikram, A. Kumar and M. Rizwan, "A Novel Fuzzy based Intelligent Demand Side Management for Automated Load Scheduling," 2020 4th International Conference on Green Energy and Applications (ICGEA), Singapore, Singapore, 2020, pp. 171-175.

[4] Pedro Faria, Tiago Soares, Zita Vale and Hugo Morais. 2014. Distributed generation and demand response dispatch for a virtual power player energy and reserve provision. *Renewable Energy* 66 (2014), pp. 686-695.

[5] T. Logenthiran, D. Srinivasan, and T.Z. Shun. 2012. Demand Side Management in Smart Grid Using Heuristic Optimization. *IEEE Transaction on Smart Grid 3*, 3 (Sept 2012), pp. 1244-1252.

[6] João Soares, Mohammad Ali, Fotouhi Ghazvini, Marco Silva and Zita Vale. 2016. Multi-dimensional signalling method for population-based metaheuristics: Solving the large-scale scheduling problem in smart grids. *Swarm and Evolutionary Computation* 29 (2016), pp. 13 – 32.

[7] Anupam Trivedi, Dipti Srinivasan, Subhodip Biswas, and Thomas Reindl. 2016. A genetic algorithm – differential evolution based hybrid framework: Case study on unit commitment scheduling problem. *Information Sciences* 354 (2016), pp. 275-300.

[8] D. S. Kumar, D. Srinivasan, and T. Reindl. 2016. A novel hybrid method for planning and allocation of DGs in medium voltage networks considering voltage regulation and relay coordination. In *IEEE Congress on Evolutionary Computation (CEC)*. Pp. 3654–3661.

[9] Zhongping Wan, Guangmin Wang, and Bin Sun. 2013. A hybrid intelligent algorithm by combining particle swarm optimization with chaos searching technique for solving nonlinear bilevel programming problems. *Swarm and Evolutionary Computation* 8 (2013), pp. 26–32.

[10] Swagatam Das and Ponnuthurai Nagaratnam Suganthan. 2011. Differential evolution: A survey of the state-of-the-art. *IEEE transactions on evolutionary computation* 15, 1 (2011), pp. 4–31.

[11] L Lakshminarasimman, S Subramanian, "Short-term scheduling of hydrothermal power system with cascaded reservoirs by using modified differential evolution," in IEEE Proceedings-Generation, Transmission and Distribution, vol. 153, no. 6, pp. 693-700, November 2006

IOP Conf. Series: Materials Science and Engineering **946** (2020) 012008 doi:10.1088/1757-899X/946/1/012008

Molecular Complexity of Working Fluids Dedicated to Organic Rankine Cycle (ORC)

Dominika Matuszewska

AGH University of Science and Technology, A. Mickiewicza Av. 30, 30-059 Krakow, Poland

dommat@agh.edu.pl

Abstract. Organic Rankine Cycle (ORC) power plants are the most well-known and mature technology used to converting low and medium- temperature heat into electricity, especially from waste heat and renewable energy sources (including geothermal and solar thermal). Despite the level of development of ORC systems available on the market, there are still many aspects in the context of improving those units – constructing economically profitable low power unit, improving efficiency, selection an appropriate working fluid, etc. This paper analyzes 17 workings fluids dedicated to ORC system and discusses an existence of low stability region in their dense vapors and its influence on some peculiarities in behaviors of selected isentropic and dry fluids. The analysis shows relation between the retrograde phenomena in the flow of ORC fluids and mechanical and thermodynamic stability parameters.

1. Introduction

In opposition to fossil fuel, heat from waste or renewable resources is mostly available from low to medium temperature (in the range from 70 to 350 °C) [1]. The use of Organic Rankine Cycles (ORCs) allows for exploitation of heat that would be wasted in other cases. The working principle of Organic Rankine Cycle (ORC) power plan is common to the most widely used process for power generation – the Clausius-Rankine Cycle. The main difference between those two cycles is the use of organic working fluid in ORC instead of water (steam), which is used in conventional Clausius-Rankine Cycle. The use of organic substances with a lower boiling point and a higher vapor pressure instead of water allows to utilized low and medium temperature heat sources to produce electricity. The choice of working fluid is one of the most crucial things in designing ORC system – there is a lot of research [2-4], which discuss working fluid selection for ORC power plant. Authors agree that an organic working fluid selected to ORC should be fitted to heat source and its thermodynamic properties, thanks to that higher efficiencies of both cycle and expander can be obtained [5,6].

Maizza and Maizza [7] showed that there are other important features in working fluid selection for ORC such as: low toxicity, good compatibility and chemical stability in operation with other materials, low flammability, corrosivity and small potential for decomposition. Within the last years, many novel working fluids have been introduced, such as hydrofluoroolefines (HFO). A new generation of working fluids meet the criteria of environmental friendliness and are characterized by significantly smaller Global Warming Potential (GWP) compared to the currently widely used refrigerants [8]. Also, it can be observed an increase of siloxanes based on Organic Rankine Cycles across the Europe. They are increasingly chosen because of their advantages, such as: long-lasting, inflammable and more stable than carbo-organic fluids [9].

Content from this work may be used under the terms of the Creative Commons Attribution 3.0 licence. Any further distribution of this work must maintain attribution to the author(s) and the title of the work, journal citation and DOI.
Published under licence by IOP Publishing Ltd

Lukawski, Tester and DiPippo [10] in their work analyzed impact of molecular structure of working fluids on Organic Rankine Cycle plant efficiency. They showed that maximum utilization efficiency of ORC system and the temperature of heat source can be predicted by using two parameters: the reduced ideal gas capacity and the critical temperature of working fluid. Works [11] and [12] analyzed an influence of molecular complexity and low-stability region on dense gas flows in ORC fluids. Evaluation of Using Gas Turbine to Increase Efficiency of the Organic Rankine Cycle was obtained by Matuszewska and Olczak [13].

2. Characteristic of ORC fluids

The organic fluids dedicated to ORC system can be used at much lower evaporation temperature than in water-steam Clausius-Rankine Cycle, what is also important – their superheating is not required. In order to generate mechanical power, organic fluids are expanded in turbine or expander. The dynamic of expansion strongly depends on the process conditions including the initial thermodynamic state and expansion ratio [11]. The working fluid selection is driven by different requirements and is strongly related to specific ORC application (temperature of heat source, compatibility, flammability and toxicity, the environment friendliness). Mostly ORC fluids can be characterized by belonging to organic chemical groups, high molecular weight, much lower temperature and pressure than water and the slope shape, which cause that the superheating of them is not required. These factors determinate that those working fluid can be used to the utilization of low and medium temperature heat. However, there are still some problems during ORC fluids selection, among others: an incomplete data in its thermal properties, problems with drop-in replacement of old generation and environmental harmful fluids.

On purpose of this paper, 17 working fluids have been chosen, table 1. The fluids have been arranged according to their chemical groups. In table 1, are listed representative of the formerly widely used fluids such as hydrocarbons (HC's) and their derivatives (HFC, HCFC). Nowadays, they use of theme is taken under limitation due to their harmful impact on global warming and ozone depletion (high GWP and ODP). Also, representatives of new generation organic fluids have been taken under consideration – hydrofluoroolefins (HFOs) dedicated to low temperature ORC plant and siloxanes – to medium temperature ORC systems.

Table 1. Working fluids dedicated to ORC systems and their selected properties

Group	Name	Mass molar M [g/mol]	Pc [bar]	Tc [K]	NBP [K]	Acentric Factor ω
Hydrocarbons (HCs)	R600	58.12	37.9	425.1	273.0	0.201
	Toluene	92.14	41.3	591.8	383.8	0.266
	n-dodecane	170.0	29.3	374.9	256.8	0.357
Hydrofluorocarbons (HFCs)	R134a	102.0	40.6	374.2	247.1	0.327
	R227ea	170.0	29.3	374.9	256.8	0.357
	R245fa	134.1	36.5	427.2	288.3	0.378
Hydrofluoroolefins (HFOs)	R1234yf	114.0	33.8	367.9	243.7	0.276
	R1234zeE	114.0	36.4	382.5	254.2	0.313
	R1234zeZ	114.0	35.3	423.3	282.9	0.327
Hydrochlorofluorocarbons (HCFCs)	R123	152.9	36.6	456.8	301.0	0.282
Siloxanes	MM	162.4	19.4	518.8	373.4	0.418
	MDM	263.5	14,2	564.1	425.7	0.529
	MD2M	310.7	12.3	599.4	467.5	0.668
	MD4M	459.0	8.8	653.2	533.9	0.825
	D4	296.6	13.3	586.5	448.5	0.592
	D5	370.8	11.6	619.2	484.1	0.658
	D6	444.9	9.6	645.8	518.1	0.736

The organic fluids from table 1 can be divided into two groups: the first one with high-boiling point (NBP) close to the ambient temperature (include siloxanes, n-dodecane and toluene), the second one with low-boiling point, which includes almost all hydrocarbons (and their derivatives). Table 1 shows also correlation between growth of acentric factor ω and rise of molecular mass of organic fluid.

2.1. The saturation vapor curve

Thanks to the shape of the saturation line, organic fluids can be divided into three groups: wet, isentropic and dry. The shape of the saturation line is one of the most important aspects characteristic ORC plant – it is taken under consideration during system designing and equipment selection – because it influences system efficiency. The figure 1 represents all three saturation vapor curve types: a negative slope for wet fluid (for water ammonia), nearly vertical vapor saturation curve for isentropic fluid (R11) and a positive slope for dry fluid.

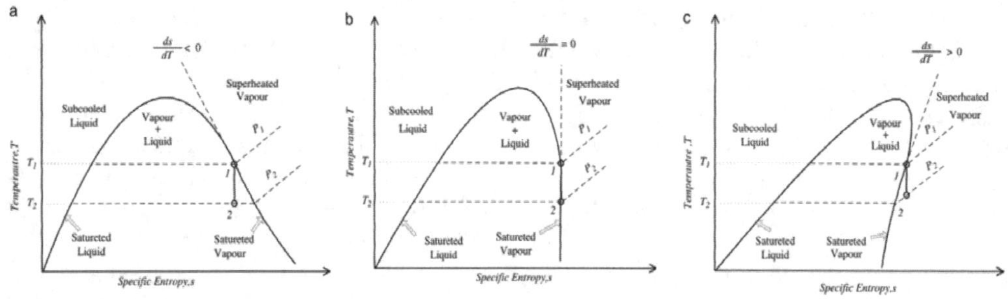

Figure 1. The saturated vapor curve for fluids (a) wet, (b) isentropic and (c) dry [3]

Dry and isentropic fluids are best recommended for ORC plant – both do not need superheating, which eliminates concerns connected with turbine blades corrosion. From these two, the isentropic fluid is more recommended– because there is no need to use regenerator in such system and the saturated vapor at the turbine inlet expands along vertical line and remain saturated leaving the turbine.

2.2. ORC fluids behavior on saturated line

Recent research focuses on the influence of organic fluids molecular characteristics and dense-gas flow phenomena, which is not well known nowadays. Heavy organic fluids dedicated to ORC systems have composed molecular structure which causes that they do not behave classical near the saturated vapor curve (x=1). In this light, it seems very important to study and analyze the measures of fluid complexity and associated theme with fluid behaviour and parameters.

From the molecular complexity, the shape of slope of saturation vapor curve can be predicted. Thereby, the information whether fluid is wet, isentropic or dry can be given. For this purpose, coefficient σ can be used – this factor is usually referred to the reduced temperature Tr = 0.7 and it defined by equation 1:

$$\sigma_{Tr=0.7} = \frac{T_c}{R}\left(\frac{ds}{dT}\right)_{x=1} = \frac{T_c}{R}\left[\left(\frac{\partial s}{\partial p}\right)_T\left(\frac{dp}{dT}\right)_{x=1} + \left(\frac{\partial s}{\partial T}\right)_p\right]_{Tr=0.7} \quad (1)$$

Equation 1 can be presented in simplify form:

$$\sigma_{Tr=0.7} = \frac{T_c}{R}\left[\frac{c_p^0}{T} - \frac{R}{p}\left(\frac{dp}{dT}\right)_{x=1}\right]_{Tr=0.7} \quad (2)$$

The σ coefficient is a heat capacity function and is directly related to the fluid molecular structure. When molecular complexity increase, the decrease of the heat capacity ratio γ, can be observed. It tends the slope of the saturation vapor curve become positive – the more positive slope is, then the more complex molecular structure can be observed (figure 2):

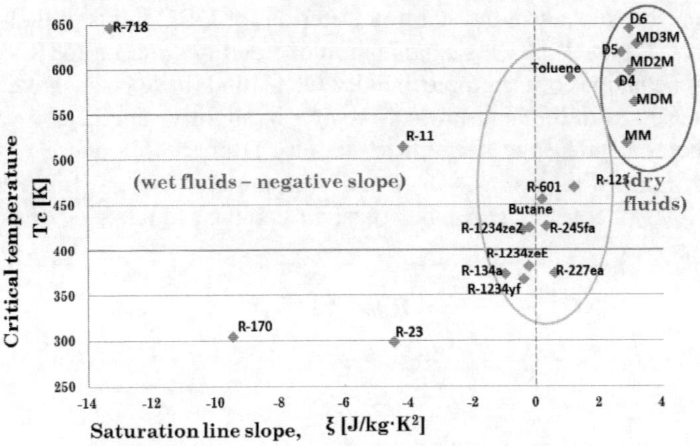

Figure 2. Wet, isentropic and dry fluids [12]

The 'retrogradicity' parameter proposed by Lambrakis can be used as counterpart formulation [14]:

$$r_\sigma = \left(\frac{\partial T}{\partial v}\right)_p \left(\frac{ds}{dp}\right)_{x=1} \geq 0 \tag{3}$$

This parameter is a very useful tool during supercritical and high-pressure subcritical ORC cycle analysis to shows difference from normal fluid behaviour. Because this case considered as real-gas expansion in the expander, temperature and pressure deeply influence variation of all physical constants. Eq. 3 can be written in form:

$$r_\sigma = \left(\frac{c_p^{"}T}{\Delta h_{lv}}\right)\left(\frac{z_{vT}}{z_{Tv}}\right)^{"}\left(1-\frac{v'}{v"}\right) - 1 \geq 0 \tag{4}$$

Where parameters z_{Tv} and z_{vT} are two Virial Compressibility Derivatives (VCD) functions [15,19]. The use of Maxwell thermodynamic relations and the Legendre transformation can help to obtain VCD's:

$$z_{Tv} = z + T\left(\frac{\partial z}{\partial T}\right)_v = z + T_r\left(\frac{\partial z}{\partial T_r}\right)_{vr} \tag{5}$$

$$z_{vT} = z - v\left(\frac{\partial z}{\partial v}\right)_T = z - v_r\left(\frac{\partial z}{\partial v_r}\right)_{Tr}$$

where $z = pv/(RT)$ is the compressibility factor. In eq. 4, first term correspondent to the reciprocal of Jacob's Number Ja and is always bigger than one, in the same time z_{vT}/z_{Tv} ratio is smaller than one. There are some cases when the multiplication of these two components can gives value close to one. However, considering that the specific volume ratio v'/v" is negligible at the sub-critical pressure are, the retrogradicity parameter can be less than zero.

Analyze shows that the stability conditions have significant influence on organic fluid in are close to the critical point and saturation line (dense gas region). It is crucial to considered whether fluid properties respond to that. Below is presented the thermodynamic stability condition for one-component homogeneous pvT system:

$$D = \frac{\partial(T,-p)}{\partial(s,v)} = \begin{vmatrix} (\partial T/\partial s)_v & (\partial T/\partial v)_s \\ (\partial T/\partial v)_s & (\partial p/\partial v)_s \end{vmatrix} \geq 0 \tag{6}$$

With some changes, it can be presented as:

$$D = -\left(\frac{\partial p}{\partial v}\right)_s \left(\frac{\partial T}{\partial s}\right)_p = -\frac{T}{c_v}\left(\frac{\partial p}{\partial v}\right)_T \tag{7}$$

The reduced form can be written:

$$\bar{D} = D\left(\frac{v_c}{T_c}\right)^2 = \frac{R}{c_v} z_{vT}\left(\frac{T_r}{v_r}\right)^2 \tag{8}$$

This equation can be used to analyze the stability criterion for ORC fluids in their dense vapor region on the saturation curve. Figure 3 presents examination of stability criterion for R245fa.
Some peculiarities in behaviour of isentropic and dry ORC fluid, have been shown during observation. A low stability of dense gas in the high-pressure region of slightly superheated vapor can be noticed. From figure 3 can be concluded that the reduce stability D of R245fa dense vapor shows in region with the temperature range $438 < T_r < 445$ K and pressure $0.85 < p_r < 1.3$. The local minima and saddle points of isoclines of constant D can be reveal around the saturated vapor state.

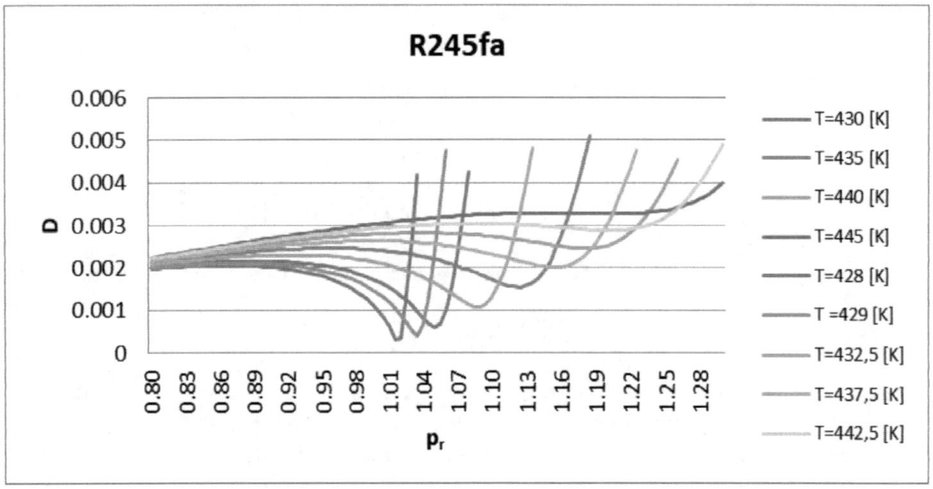

Figure 3. Reduce thermodynamic stability D of R245fa

The characteristic gas parameter is its isentropic index k_s, which shows the relation below:

$$k_s = \frac{\rho}{p}\left(\frac{\partial p}{\partial \rho}\right)_s = -\frac{\hat{V}}{p}\left(\frac{\partial p}{\partial \hat{V}}\right)_s = \gamma \frac{z_v}{z} \tag{9}$$

Figure 4 shows an example of the relationship of the isentropic index as a function of reduced pressure and temperature for R245fa:

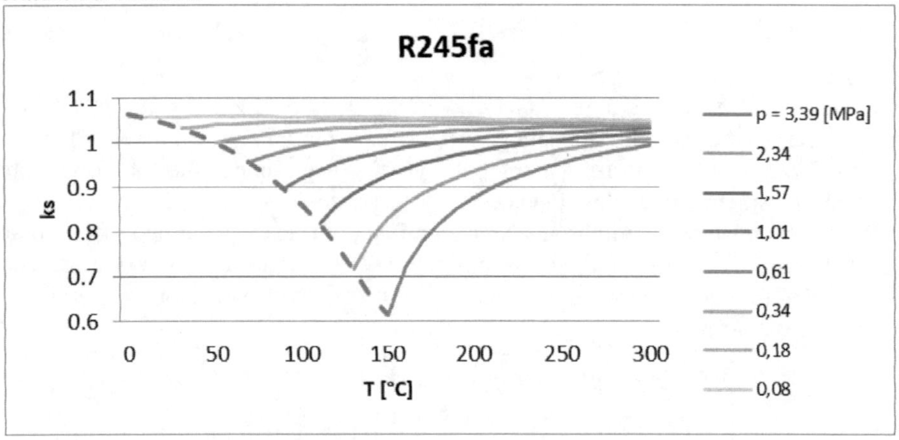

Figure 4. Isentropic index for R245fa

Figure 4 clearly shows that the isentropic index ceases to be sensitive to pressure changes as the temperature increase. Its largest fluctuations are observed near the saturation curve, and its value is inversely proportional to the pressure increase.
Reduced thermodynamic stability criterion D on the saturated vapor line (x=1)

for all selected working fluids has been shown on figure 5. The working fluid arrangement is based on the acentric factor ω. This listing has been made for three important points on the saturated vapor curve: including temperature corresponded to s” =max and h”=max and reduced temperature Tr=0.7.

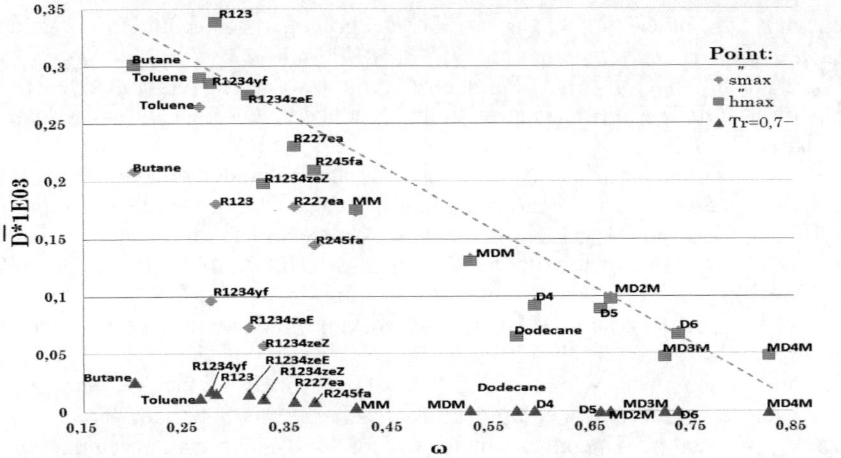

Figure 5. Stability criterion D vs acentric factor ω for ORC fluids

Analyzing fig. 5, it can be noticed that the stability criterion D on the saturation line for reduced temperature Tr=0.7 achieved values close to zero for high molecular mass and complex fluids (siloxanes). The stability criterion (for reduced temperature Tr=0.7) for hydrocarbons and its derivatives is minimally higher and positive. Almost linear dependence to the Pitzer’ s criterion ω obtain the stability criterion D for temperature corresponding to h''_{max}. There is hard to observe any regularity to the ω for s''_{max} data. The stability criterion D is almost the same for both h''_{max} and s''_{max} conditions on the saturated vapor line for siloxanes. Also, for all considered ORC fluids, the stability criterion D for temperature corresponding to s''_{max} is lower to those with h''_{max}.

3. Conclusions

The analysis shown in this paper confirms that some peculiarities appear in high dense gas region of ORC fluids. They are corresponding to the thermodynamic parameters and compressible flow conditions. For high -molecular mass and complex fluids such as siloxanes, it has been observed that the stability criterion D on the vapor-liquid line is nearly zero. Higher and positive value of stability criterion ($D = 0.14 \div 0.23 \cdot 10^{-3}$) is obtained for butane, toluene, R123, R227ea, R245fa and the HFO group R1234ze. It is easy to observe nearly linear dependence of D respect to ω factor for the conditions of enthalpy maxima h''_{max}. However, the maxima of saturated vapor entropy s''_{max} for all investigated ORC fluids do not show any regular variation to the Pitzer's criterion ω. It should be noticed that the s''_{max} maxima for all investigated ORC fluids corresponds to the lower reduced temperatures than values obtained for the h''_{max}.

The working fluid selection for ORC cycles is hard and complex – a lot of factors and relationships occurred that should be always individually considered during designing process. The complexity of ORC fluids selection is hard to understand good enough due to the lack of thermodynamic data. The best working fluids for ORC are isentropic and dry ones but they should be chosen individually, depending on temperature of heat sources, environmental criteria and thermodynamic properties of ORC fluids. It is important to remember that all fluids dedicated to ORC systems have high-molecular complexity which influences their behavior. The use of complete real properties of ORC fluids can improve selection process by showing real performances and efficiency of ORC systems.

Acknowledgements

This work was supported by AGH – University of Science and Technology (Project 16.16.210.476).

References

[1] Mikielewicz, D., Mikielewicz J., Criteria for selection of working fluid in low-temperature ORC. *Chemical and Process Engineering,* 2016; 37 (3): 429-440
[2] Dai, Y., J. Wang, and L. Gao, Parametric optimization and comparative study of organic Rankine cycle (ORC) for low grade waste heat recovery. *Energy Conversion and Management*, 2009; 50(3): 576-582.
[3] Bao, J. and L. Zhao, A review of working fluid and expander selections for organic Rankine cycle. *Renewable and Sustainable Energy Reviews*, 2013; 24: 325-342.
[4] Chen, H.,D.Y. Goswami, and E.K. Stefanakos, A review of thermodynamic cycles and working fluids for the conversion of low-grade heat. *Renewable and Sustainable Energy Reviews*, 2010; 14(9): 3059-3067.
[5] Liu, B.T., Chien, K.H., Wang, C.C. Effect of working fluids on organic Rankine cycle for waste heat recovery. *Energy*, 2004; 29: 1207–1217
[6] Nowak, W., Borsukiewicz-Gozdur, A., Stachel, A.A., Using the low-temperature Clausius–Rankine cycle to cool technical equipment. *Appl. Energy*, 2008; 85: 582–588
[7] Maizza V., Maizza A., Unconventional working fluids in organic Rankine-cycles for waste energy recovery systems. *Appl. Therm. Eng.*, 2001; 21: 381-390.
[8] Eyerer, S., Dawo, F., Kaindl, J., Wieland, C., & Spliethoff, H. Experimental investigation of modern ORC working fluids R1224yd(Z) and R1233zd(E) as replacements for R245fa. *Applied Energy,* 2019; 240: 946–963
[9] Erhart, T., Gölz, J., Eicker,U., van den Broek, M., Working Fluid Stability in Large-Scale Organic Rankine Cycle-Units Using Siloxanes—Long-Term Experiences and Fluid Recycling. *Energies*, 2016; 9: 422
[10] Lukawski, M.Z., Tester, J.W., DiPippo, D., Impact of molecular structure of working fluids on performance of organic Rankine cycles (ORCs). *Sustainable Energy & Fuels* 2017; **1**: 1098-1111
[11] Harinck, J., Guardone, A., Colonna, P., The influence of molecular complexity on expanding flows of ideal and dense gases. *Physics of Fluids*, 2009; 21(8): 086101
[12] Matuszewska, D., Sztekler, K., Górski, J., An influence of low-stability region on dense gas phenomena and their peculiarities in the ORC fluids. *MATEC Web of Conferences*, 2014; 18: 03005
[13] Matuszewska, D. and P. Olczak, Evaluation of Using Gas Turbine to Increase Efficiency of the Organic Rankine Cycle (ORC), *Energies*, 2020, 13(6): 1499
[14] Colonna, P., Guardone, A., Nanna, N.R., Siloxanes: a new class of candidate Bethe-Zel'dovich-Thompson fluids. *Phys. Fluids*, 2007; 19: 086102

IOP Conf. Series: Materials Science and Engineering **946** (2020) 012009 doi:10.1088/1757-899X/946/1/012009

Kinetics study of methane production from anaerobic digestion of sludge and wastewater recycled pulp and paper

M. Bakraoui[1,*], F. Karouach[a], B. Ouhammou[a], N. Lahboubi[b], Y. El Gnaoui[a], O. Kerrou[a], M. Aggour[a], H. El Bari[a]

[a] Renewable Energy and Environment laboratory, Faculty of Sciences-Kenitra, Ibn Tofail University, Morocco

[b] Process Engineering, Polymer and agro-resources Laboratory, Faculty of Sciences-Kenitra, Ibn Tofail University, Morocco

E-mail address: mohammed.bakraoui@uit.ac.ma

Abstract. Methane yield is one of the important factors that indicate the anaerobic digestion (AD) process efficiency. The AD is an available treatment of recycled pulp and paper sludge (RPPS) and recycled pulp and paper wastewater (RPPW). This process protects the environment and produces methane, which is considered as renewable energy. In this study, the anaerobic treatment of the RPPW and RPPS were studied using 1L completely stirred tank reactor (CSTR) at laboratory scale with mesophilic conditions. However, a kinetic study was carried out using four models in order to compare and adjust the experimental results. The experimental results show that the RPPW and RPPS methane productions were between 691 and 945 mL/g based on volatile solid (VS), and the corresponding digestion time was about 26 days for the both experiment. The corresponding digestion time could be understood through the methane production kinetic models by digestion experiments within a short time. Different kinetics models were tested including Richards, Logistic, modified Gompertz and Cone models. The obtained results show the modified Gompertz model as the suitable model since it gives the lowest deviation from experimental results. Therefore, the proposed model could be used to accurately predict behavior of AD processes.

1. Introduction

The AD is a recommended technology that can be applied to reduce the pollution, by the reduction of organic matter, and to produce methane as a renewable energy that we can exploit in different way [1, 2]. The recycled pulp and paper is a growing industry which plays a significant part in the growth of the economy in the world [2, 3]. However this industrial sector consumes a significant amount of resources, raw materials and energy. Furthermore, it produces a considerable amount of organic waste, which is not suitable for the production of new paper [4]. Correspondingly, the biodegradable organic fractions and hydrolysis rates of substrates are two important parameters for anaerobic biodegradation. In fact, these parameters determine, to a certain extent, both the design and economic details of biomethane plants. A huge quantity of the sludge produced by paper mills is considered as one of the most serious environmental problems [5]. In general term, we use the anaerobic batch tests as standard method for the determination of methane yield potential [6]. Structured kinetic models are used for the simulation of biomethane production from anaerobic treatment by biodegradation of RPPS and RPPW to produce renewable energy. The structured models include multiple stages of the material degradation in CSTR reactors which have made it possible to predict the actual response of the process.

Content from this work may be used under the terms of the Creative Commons Attribution 3.0 licence. Any further distribution of this work must maintain attribution to the author(s) and the title of the work, journal citation and DOI.
Published under licence by IOP Publishing Ltd

[7,8]. Compared to the Richards, Logistic, Gompertz and Cone model, the modified Gompertz model was commonly used as it was correlated with four biochemical reaction parameters, A, μ_m, λ and R. The Kinetic study of mesophilic AD show that the corresponding equations could be used for design CSTR for digesting organic wastes from pulp and paper industries, food processing industries, treatment plants wastewater [9,3].

The kinetic model of methane production by AD is a practical way to understand the ultimate technique of methane production in the standard condition [3]. In the literature, several kinetic studies of the AD process show an efficient application for monitoring and controlling the stability parameters, using different kinds of reactors and substrates. The kinetic model equations could be used to predict dynamic simulation of the AD on the principle of a complex matrix series for constant kinetic values [9].

The purpose of this paper is to apply kinetic models on the AD process of RPPS and RPPW. The digester used in this study is a CSTR on batch mode. A number of modified kinetic models as Richards, Logistic, Gompertz and Cone, have been used to evaluate their applicability on experimental results of methane production of RPPW and RPPS experiments. In addition, the purpose of this paper is to describe the kinetics parameters of methane production.

2. Material and methods

2.1. Inoculum characterization

Starting the process of AD requires the use of an inoculum rich in bacteria that will degrade the organic matter. The inoculum improves the production of biogas; at the same time it reduces the retention time. The inoculum used in all experiments comes from a mesophilic anaerobic digester treating sludge from a wastewater treatment plant in Morocco. Table 1 shows its main physicochemical characteristics. Initially, the digester was fed with 7g VS / L of inoculum, which is equivalent to 342 mL.

2.2. Substrate characterization

The sludge and the wastewater used in the experiment were recovered from a recycled pulp and paper manufactory in Morocco [10,1]. The sludge was kept after gravitational static decantation, at laboratory scale, in order to eliminate the supernatant and recover only the deposit of the effluent "sludge". Inoculum, sludge, and wastewater analysis were performed according to the "Standard Methods for the Examination of Water and Wastewater" (12). Indeed, recent values show that the COD of pulp and paper effluent corresponds to values in the range of (3770 - 9330mg L^{-1}) and BOD_5 in the range of (816-2495 mg L^{-1}) [11].

Table 1. Substrates and inoculums characteristics.

Parameters	Unit	Inoculum	Wastewater recycled pulp and paper	Sludge recycled pulp and paper
pH	--	7.2 ± 0,11	7.5	7.1
COD	mg/l	7741.12 ± 140	7910.5	9230
BOD_5	mg/l	--	4430.5	6100
COD/BOD_5	--	--	1.51	1.94
Alk	$mgCaCO_3/l$	2750	2250	1875
TS	g/l	29.05 ± 0.63	55.70	105.07
VS	g/l	9.71 ± 0.46	35.25	55.22
MS	g/l	19.339 ± 0.43	20.457	49.855
VS (% TS)	%	33.42	63.28	52.55

IOP Conf. Series: Materials Science and Engineering **946** (2020) 012009 doi:10.1088/1757-899X/946/1/012009

2.3. The anaerobic digester

The reactor used in the laboratory for AD, consisted of 1L Pyrex CSTR (Continuous Stirred-tank reactor), equipped with magnetic stirring. It worked in batch mode. This reactor is characterized by four orifices, the first to insert the substrate, the second one for the ventilation of biogas, the third for inert gas injection (nitrogen) to maintain anaerobic conditions, and the last one to remove effluents. The schematic diagram of experimental set up is shown in Fig. 1. [12,3].

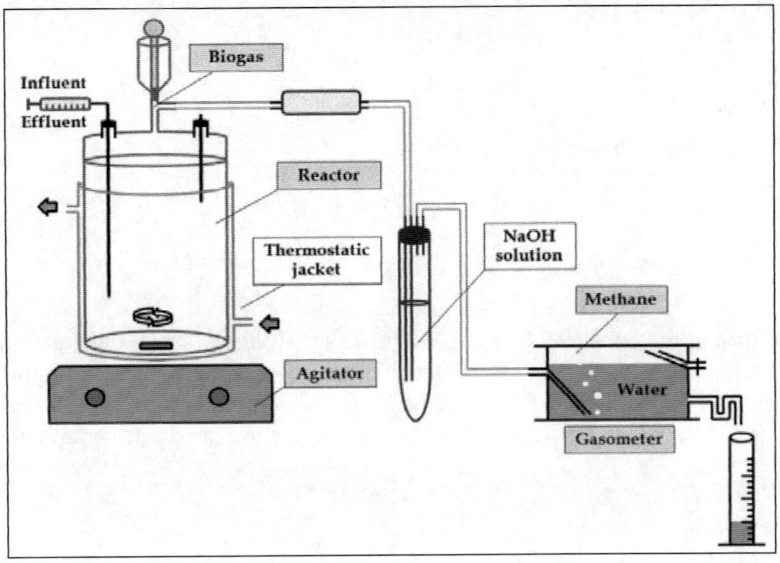

Figure 1. Schematic diagram of CSTR reactor system.

The temperature of reactor was maintained by means of a thermostatic jacket containing water at 37°C for mesophilic conditions. The volume of methane produced during the digestion process was measured by using 1L Boyle-Mariotte reservoir connected to the reactor. To remove the CO_2 produced during the process, tightly closed bubblers containing a NaOH solution (6 N) were connected between the two elements. It is necessary sometimes to close the pipe of the NaOH solution that links the reservoir to the reactor. The methane volume produced can be determined from the volume of water ejected from the reservoir (volume corrected in STP conditions: 0°C and 1 atm).

2.4. Methane measurement

The produced methane is measured by the displacement of water using a cylindrical gas-meter which capacity is one liter linked to a graduated tube of the same volume. The volume of water in the tube is equivalent to the methane produced. The cumulative methane volume is standardized with taking account the effect of atmospheric pressure, the water vapor temperature, and the pressure of the room at the time of measurement, the results are expressed under normal conditions (0 ° C at 760 mm Hg) [13].

The temperature is determined by means of the thermometer, the pressure is considered as constant throughout the process, which is equal to the atmospheric pressure. From the law of perfect gases, expressed under normal and experimental conditions:

At normal condition:

$$P_N V_N = nRT_N$$

With:

$$K = \frac{P_N V_N}{T_N} \quad (1)$$

At ambient conditions:

$$P_R V_R = nRT_R$$

With:

$$K = \frac{P_R V_R}{T_R} \quad (2)$$

The relation (1) and (2) we have:

$$\frac{P_R V_R}{T_R} = \frac{P_N V_N}{T_N} \quad (3)$$

The final formula obtained to be applied is:

$$V_N = V_R \frac{(P_{atm} - P_v)}{P_N} \frac{T_N}{T_R} \quad (4)$$

With:

V_N: Normal volume of methane (0 ° C, 1 atm) (NmL), V_R: Volume of real methane (mL), P_N: Pressure at normal conditions (0 ° C, 1atm), P_R: Actual pressure at experimental conditions (mbar), P_{atm}: Atmospheric pressure in mmHg. P: Vapor pressure at room temperature mmHg, T_N: Temperature at normal conditions (0 ° C (273 K), 1atm), T_R: Actual temperature at experimental conditions (° C (K)) [14].

2.5. Sigmoidal models

A number of sigmoidal models are found in the literature; such as the Gompertz, Richards, Standard and Logistic models [14,15]. Modeling microbial growth can evaluate several parameters; such as specific growth rate and latency. The bacterial growth curve showed the minimal phase (in our case zero) and the enhancement of the maximum rate of growth (μ_m) in the case named as lag phase (λ) [16]. Thus, when the Bacterial growth reaches its limit, this shows that the growth curve reaches the maximum limit A [16].

The following curve shows three essential parameters for understanding the biological process of organic matter degradation. μ_m presents the tangent of the linear curve during methane production; λ shows the intersection of the curve with the x-axis, and a shows the intersection of the curve with the y-axis (also presents the maximum cumulative methane).

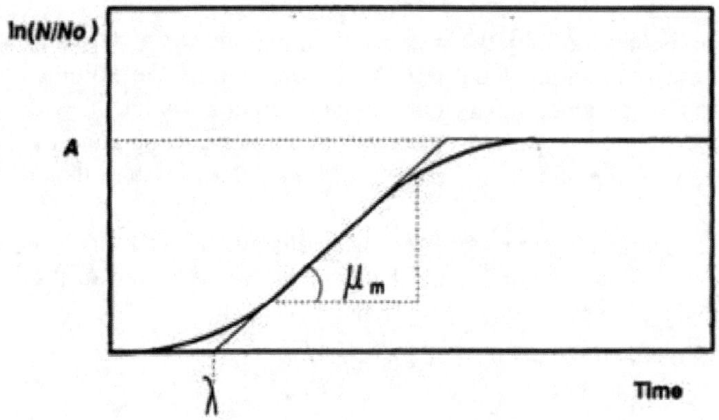

Figure 2. Typical curve of methane production

IOP Conf. Series: Materials Science and Engineering **946** (2020) 012009 doi:10.1088/1757-899X/946/1/012009

Table 2. Models use in bacterial growth and their modified forms according to Zwietering et al. [13].

Model	Equation	Modified equation
Logistic	$y=\frac{a}{\left[1+a\exp(b-cx)\right]}$	$y=\frac{A}{\left[1+\exp\left(\frac{4\mu_m}{A}(\lambda+t)+2\right)\right]}$
Gompertz	$y=a\exp\left[-\exp(b-cx)\right]$	$y=A\exp\left\{-\exp\left[\frac{\mu_m.e}{A}(\lambda-t)+1\right]\right\}$
Richards	$y=a\left\{1+v\exp\left[k(\tau-x)\right]\right\}^{\left(-\frac{1}{v}\right)}$	$y=A\left\{1+v\exp(1+v).\exp\left[\frac{\mu_m}{A}(1+v)\left(1+\frac{1}{v}\right).(\lambda-t)\right]\right\}^{\left(-\frac{1}{v}\right)}$
Stannard	$y=a\left\{1+\exp\left[-\frac{(l+kx)}{p}\right]\right\}^{-p}$	$y=A\left\{1+V\exp(1+V).\exp\left[\frac{\mu_m}{A}(1+V)\left(1+\frac{1}{V}\right).(\lambda-t)\right]\right\}^{\left(-\frac{1}{v}\right)}$
Cone	$y=\frac{a}{\left[1+(kx)^{-n}\right]}$	$y=\frac{A}{\left[1+(kt)^{-n}\right]}$

Y: cumulative specific methane production ($mLCH_4$ g VS/L), A: maximum specific methane production potential ($mLCH_4$/gVS), µm: max. Specific methane production rate ($mLCH_4$/gVS.d), e: Exp (1), λ: lag-phase (days), t: incubation time (days), v: shape coefficient.

These two equations are used from the kinetics study of author Dong Li, Kinetics of methane production and hydrolysis in AD of corn stover, to give a relationship between methane production and organic matter degradation in terms of volatile solids [17].

$$y=A\exp\left\{-\exp\left[\frac{\mu_m.e}{A}(\lambda-t)+1\right]\right\} \quad (5)$$

$$VS_{T,t}=VS_{NB}+\left(VS_{T,0}-VS_{NB}\right)*\exp\left(-K_{H,B}\times t\right) \quad (6)$$

Where $VS_{T,t}$ is the total VS content (g/L), VS_{NB} is the non-biodegradable VS content (g/L, or % VS_T), $VS_{T,0}$ is the initial total VS.

Adjustment of the pairs of experimental data (t, VS_T,t) by non- linear regression allowed the calculation of the fraction of VS_{NB} and $K_{H,B}$.

3. Results and discussion

In order to characterize the kinetics of the two experiments, and thus to facilitate the comparison between the different kinetic models applied to the production of methane; the previous models were used to adjust the experimental data. Figure 3 shows the variation of the accumulated methane volume as a function of time for the two types of the treated substrate. The results show that methane production increases with increasing time. This phenomenon is due to the organic matter degradation, which emerges in form of biogas in 26 days. This is equal to the substrate degradation time for both waste types. The methane yields in this study were higher than that of previous studies [18]. The methane yield increased in 10 days from 0.003 to 0.09 L CH_4/g COD.

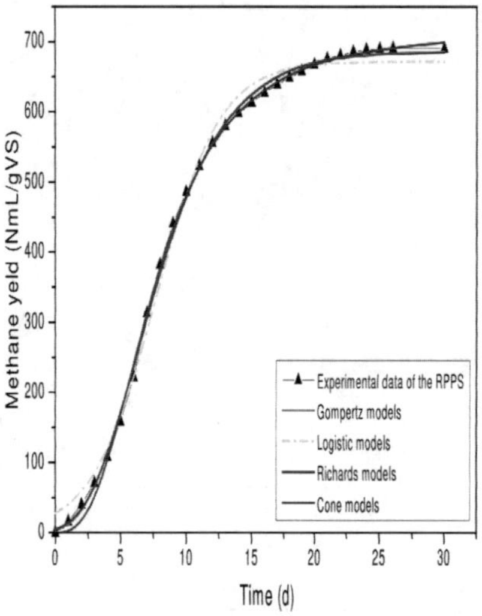

(a)

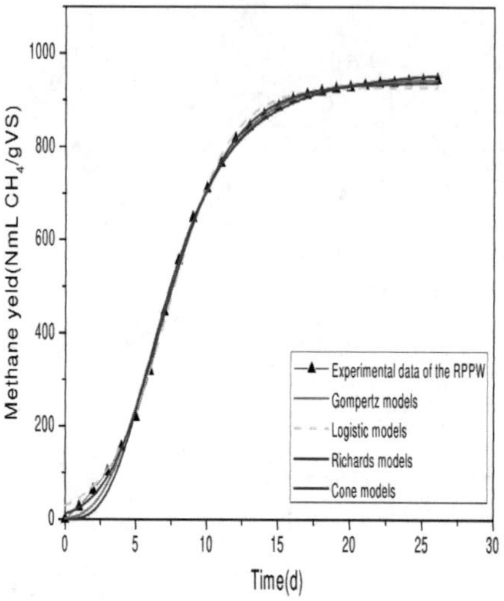

(b)

Figure 3. Cumulative methane production of: (a) RPPS, (b) RPPW.

The volume of methane produced is an important parameter used for the control and monitoring of the AD process. Indeed, a substantial production of methane reflected the proper functioning of the digester. Figure 3 represents the total volume of methane produced for both: AD of RPPS and RPPW. The efficient method to determine the methane production for different organic matter concentrations is the kinetic study using a several models [17].

The figure shows a rapid evolution of methane production with a high rate of bacteria growth during the first 10 days, which means that the bacteria are well adapted to the substrate. This growth rate decreases during the last days for both experiments.

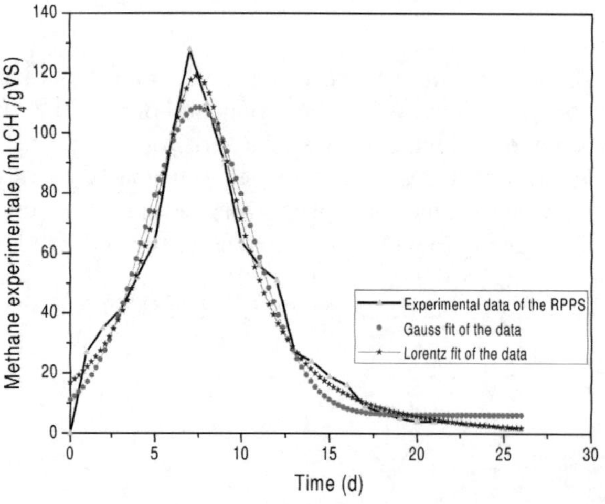

(a)

IOP Conf. Series: Materials Science and Engineering **946** (2020) 012009 doi:10.1088/1757-899X/946/1/012009

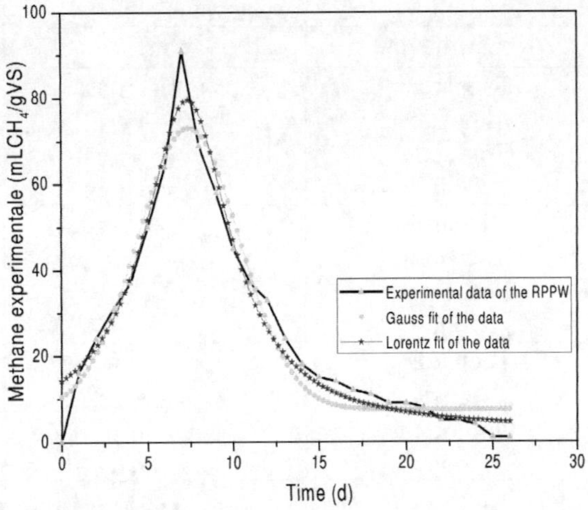

(b)

Figure 4. Experimental methane:(a) methane production rate during the AD of sludge recycled paper mill, (b) methane production rate during the AD of wastewater recycled paper mill.

The results of the simulation of methane-produced volume are illustrated in Figure 4. The results show that the simulations are in agreement with the experimental results. However, the little variations observed are related to the change in organic matter degradation and the substrate nature.

Figure 4 shows the variation of the simulated and experimental results of the produced methane volume over time. This depends on the nature, composition, and biodegradability of industrial waste.

In this study, two equations were applied for the experimental data to test which was compatible with our case study. The graph shows the experimental data and the simulation of Gauss and Lorentz equation. We deduce from the correlation vector and the variables of each equation that Lorentz is well adjusted to our study by a correlation vector of R^2 equal to 97.1%, which is higher than the one of Gauss which is equal to 94.71%. In this case, for experimental data of recycled wastewater paper mill; Lorentz equation is compatible by a correlation vector equal to 95.54% which is higher than the one of Gauss which is equal to 93.01%.

3.1. Kinetic study of methane production

Table 3 shows the values of the kinetic models used, as well as their standard deviations obtained in the two experiments. The standard deviations are respectively less than 5% and 10% for both types of waste studied. This suggests that the proposed models are well adjusted to the experimental data. The authors summarize the adjustment errors of 10% or month is a good adjustment for methane production obtained from the kinetic study by deferent model [19, 20].

The table shows the kinetic model variables applied to the experimental result. Each model gives an idea of the quantity of methane produced as a function of eliminated gVS; the volume of methane produced each day, and the hydraulic retention time λ (day).

Table 3. Kinetic parameters of average cumulative methane production curves.

Parameter	Models	Recycled paper mill wastewater	Recycled paper mill sludge
$V_0 = (mLCH_4 / gVS)$	Experimental	691	945
$A(mLCH_4 / gVS)$	Richards	685.96±3.32	937.28±3.17
	Logistic	671.4±5.8	927.9±4.5
	Gompertz	687.13±3.23	945.37±4.4
	Cone	716.37±4.99	970.8±8.3
$\mu_m(mLCH_4 / gVS.d)$	Richards	38.24±2.5	67.65±9.5
	Logistic	68.35±2.9	107.9±2.8
	Gompertz	67.4±1.3	106.71±2.23
	Cone	-	-
$\lambda(days)$	Richards	2.47±0.16	2.84±0.08
	Logistic	2.81±0.23	3.06±0.12
	Gompertz	2.5±0.1	2.75±0.1
	Cone	2.72±0.07	3.116±0.11
v	Richards	0.052	0.376
	Logistic	-	-
	Gompertz	-	-
	Cone	0.13	0.138
R^2	Richards	0.9986	0.9993
	Logistic	0.9941	0.9981
	Gompertz	0.9987	0.99865
	Cone	0.9987	0.99755
$\%error V_0 and A$	Richards	-0,73	-0,82
	Logistic	-2,84	-1,81
	Gompertz	-0,56	0,04
	Cone	3,67	2,73

From the result obtained by the kinetic study of recycled pulp and paper sludge with nonlinear regression, we observe that the theoretical methane volume varies between 927.9 to 970.8 mL CH_4/ gVS, the correlation vector is between 99.755% to 99.934%, and the time required for the degradation of organic matter is from 2.75 to 3.116 days. It can be concluded from these results that the model of Gompertz fits the experimental data because the difference between the experimental and theoretical value given by this model is very low compared to other models. In the literature indicate that the most frequently used kinetic model for methane production is the Gompertz model [21,22,23].
From the results obtained by the kinetic study of recycled pulp and paper wastewater with nonlinear regression, we observe that the theoretical methane volume varies between 695.96 to 716.37 ml CH_4/ gVS, the correlation vector is between 99.405 % to 99.876 %, and the time required for the degradation of organic matter is from 2.47 to 2.81 days. From the results, we concluded that the model of Gompertz fits the experimental data. The difference between V_0 and A (the theoretical volume) is -0.56, which is very low compared to Richards, Logistic, and Cone models. The volume of methane produced per day for each experiment of sludge and wastewater from recycled pulp and paper, respectively, are 67.65 to 106.71 mL CH_4/g VS and 38.24 to 68.35 mL CH_4/g VS.

3.2. Gompertz kinetics of hydrolysis and methane production

Fig. 5 displays both the experimental and simulated $VS_{T,t}$ and G_t according to Eq. (6) and (7). The hydrolysis kinetic model of Eq. (6) did not fit the experimental data for both the recycled pulp and paper sludge and wastewater (RPPS, RPPW). Unfortunately, the experimental value of $VS_{T,t}$ continued to fall, whereas the fitted curve leveled off. The values of VS_{NB} obtained using Eq. (6) were much greater than the experimental final VS_T. We concluded that Eq. (7) was not a good description of the entire hydrolysis process.

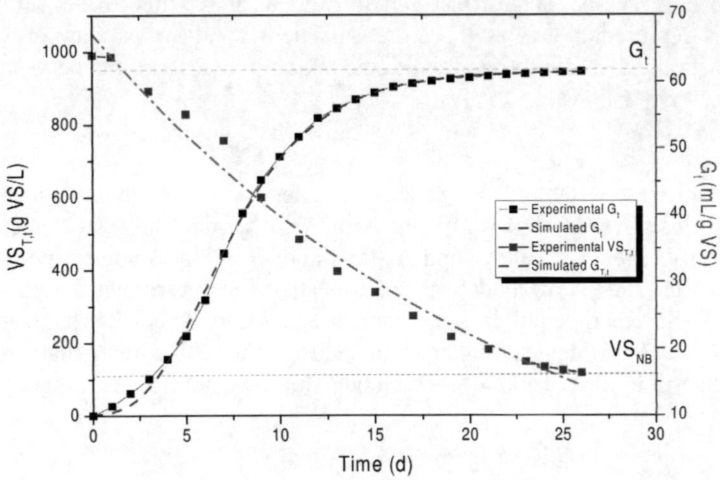

(a)

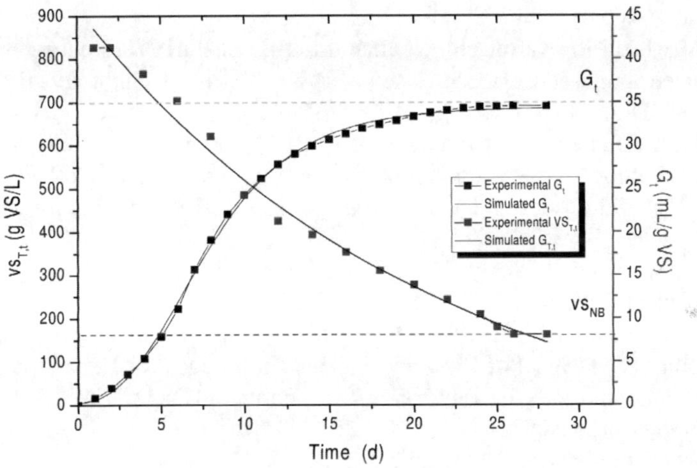

(b)

Figure 5. Gompertz kinetics of hydrolysis and methane production of: (a) RPPS, (b) RPPW

This phase, which is considered as the kinetic limiting step of the organic matter anaerobic degradation, is not a biological step strictly speaking, but a substrate reaction preparation step.
The hydrolysis rate under the best conditions depends mainly on the amount of material that remains to be hydrolyzed because the enzymes are in excess relative to the substrate. Under these conditions, its kinetics is comparable to that of a chemical-type reaction.
The hydrolysis constants determined experimentally for different types of organic matter, these constants result from analyzes of the substrate consumption or the production of VFA or even of biogas, under conditions allowing for hydrolysis kinetics determination. This is the case, for example, for methane production measurements which were carried out during the tests under controlled conditions in which all the limiting factors presented previously have been controlled and where hydrolysis proves to be effectively the same. Kinetically limiting step, that is to say in particular where

acidogenesis or methanogens are not [24]. Kinetics determined in this way has already been applied to the production of methane obtained on different fractions of RPPS and RPPW from PBM tests.
The kinetic study built on the basis of various experimental results and describing the anaerobic degradation in two successive stages, has generally made it possible to account for these different phenomena. This representation was therefore sufficient to allow a relatively fine description of anaerobic degradation under laboratory conditions, taking into account the necessary balance between the main microbial populations.

4. Conclusion

A mesophilic AD of wastewater and sludge recycled paper mill was evaluated using a 1-liter lab-scale CSTR anaerobic digesters. In this study, the Gompertz model shows a perfect fitting in kinetic modeling with a difference between V_0 and A (0.04 and -0.56) for Sludge paper mill and wastewater respectively. Modified Gompertz model provided best fit for experimental data. From the results of wastewater paper mill experimental data, we conclude that Lorentz is better adjusted to our study by a correlation vector of R^2 which is higher than Gauss. The ultimate methane production and the corresponding digestion time could only be understood thanks to the kinetic model of methane production.

5. Reference

[1] Bakraoui M, Karouach F, Ouhammou B, Aggour M, Essamri A, El Bari H,. «Biogas production from recycled paper mill wastewater by UASB digester: Optimal and mesophilic conditions,.» Biotechnology Reports 25 (2020): 0402,.
[2] Bakraoui M, Hazzi M, Karouach F, Ouhammou B, El Bari H. Experimental biogas production from recycled pulp and paper wastewater by biofilm technology Biotechnol Lett, 2019; 41:1299-1307 DOI 10.1007/s10529-019-02735-w
[3] Bakraoui M, Karouach F, Ouhammou B, Aggour M, Essamri A, El Bari H, Kinetics study of the methane production from experimental recycled pulp and paper sludge by CSTR technology, Journal of Material Cycles and Waste Management, 2019; https://doi.org/10.1007/s10163-019-00894-6,
[4] Raaz Maheshwari and al, Analysis of effluents released from recycled paper industry, Journal of Advanced Scientific Research, 2012, 3 (1): 82-85
[5] Rosazlin Ah, CheFauziahIshak, Wan RasidahKadir, and Rosenani Abu Bakar, Characterization and Feasibility Assessment of Recycled Paper Mill Sludges for Land Application in Relation to the Environment, Int J Environ Res Public Health., 2015, 12(8): 9314 9329 doi: 10.3390/ijerph120809314
[6] Adl M, Sheng K, Gharibi A. Technical assessment of bioenergy recovery from cotton stalks through anaerobic digestion process and the effects of inex- pensive pre-treatments. Appl Energ 2012;93:251e60.
[7] Batstone DJ, Keller J, Angelidaki I, Kalyuzhnyi SV, Pavlostathis SG, Rozzi A. The IWA anaerobic digestion Model no 1 (ADM1). Water SciTechnol 2002;45(10):65e73.
[8] Siegrist H, Vogt D, Garcia-Heras JL, Gujer W. Mathematical model for meso-and thermophilic anaerobic sewage sludge digestion. Environ SciTechnol 2002;36(5):1113e23.
[9] Yunqin L, Dehan W, Qing L, Lijian H. Kinetic study of mesophilic anaerobic digestion of pulp & paper sludge-biomass and bioenergy 35 (2011)
[10] Bakraoui M, El Gnaoui Y, Lahboubi N, Karouach F, El Bari H,. «Kinetic study and experimental productions of methane production from UASB reactor treating wastewater from recycled pulp and paper for the continuous test.» Biomass and Bioenergy, 2020: 139.
[11] Žarković et al., D. Žarković, M. Krgović, Lj. Rajaković.Rationalization of water consumption in paper industry, Chem Ind, 2004; 8: 327–337.
[12] B. Mohammed et al., "Effect of Thermal Pretreatment on Anaerobic Digestion Performance of Recycled Paper Sludge," 2019 7th International Renewable and Sustainable Energy Conference (IRSEC), Agadir, Morocco, 2019, pp. 1-6, doi: 10.1109/IRSEC48032.2019.9078230.
[13] Handbook of Chemistry and Physics, (2004-2005). CRC PRESS, D. R. LIDE.

[14] Altas L, Inhibitory effect of heavy metals on methane-producing anaerobic granular sludge, J. Hazard. Mater, 2009; 162: 1551-1556, http://dx.doi.org/ 10.1016/j.jhazmat.2008.06.048.

[15] Zwietering MH, Jongenburger I, Rombouts FM, Riet KV. Modelling of the bacterial growth curve, Appl. Environ.Microbiol, 1990;56: 1875-1881. DOI: 0099-2240/90/061875-07$02.00/0.

[16] Aidan W, Niamh P. Modelling methane production kinetics of complex poultry slaughterhouse wastes using sigmoidal growth functions,Renewable Energy, 2017; 104: 50-59

[17] Dong Li, Xianbo H, Qingjing W, Yuexiang Y, Zhiying Y, Zhidong L, Yajun H, Xiaofeng L. Kinetics of methane production and hydrolysis in anaerobic digestion of corn stover, Energy, 2016; 102: 1-9

[18] Haider MZ, SitiRoshayu H, NQZ , Hamidi AA, Irvan D. The start-up performance of modified anaerobic baffled reactor (MABR) for the treatment of recycled paper mill wastewater,Journal of Environmental Chemical Engineering, 2013; 61–64

[19] Raposo F, Borja R, Martín MA, Rubia MAD et Rincón B. Influence of inoculum- substrate ratio on the anaerobic digestion of sunflower oil cake in batch mode: Process stability and kinetic evaluation. Chem. Eng. J, 2009; 149: 70-77.

[20] Budiyono, Iqbal Syaichurrozi and SiswoSumardiono, Kinetic Model of Biogas Yield Production from VinasseatVariousInitialpH: Comparisonbetween Modified Gompertz Model and First Order Kinetic Model, Research Journal of Applied Sciences, Engineering and Technology, 2014; 7(13): 2798-2805,

[21] Patil JH, Raj MA, Muralidhara PL, Desai SM, Mahadeva RGK. Kinetics of anaerobic digestion of water hyacinth using poultry litter as inoculum, Int. J. Environ. Sci. Dev. 2012; 32: 94-98, http://dx.doi.org/10.7763/ IJESD.2012.V3.195.

[22] Ghatak MD, Mahanta P. Comparison of kinetic models for biogas production rate from saw dust, Carbon 2014; (3) 248-254, http://dx.doi.org/ 10.15623/ijret.2014.0307042.

[23] Bakraoui M, Karouach F, Belhadj S, Joute Y, Ouhammou B, Aggour M, Essamri A, El Bari H. Modified Gompertz kinetic study of methane production from anaerobic digestion of recycled paper mill sludge, 26th European Biomass Conference and Exhibition: 14-17 May 2018, Copenhagen, Denmark, 849 – 854p.

[24] Sanders TM, Veeken AHM, Zeeman G. & Van Lier JB. Analysis and optimisation of the anaerobic digestion of the organic fraction of municipal solid waste, chap. 3, In: MATAALVAREZ Eds. Biomethanization of the organic fraction of municipal solid waste, London : IWA publishing, 2003; 313 p.

Acknowledgments
The authors are very grateful to IRESEN (Institut de Recherche en Energie Solaire et Energies Nouvelles- www.iresen.org) for funding this research to become a Project "InnoThermo-InnoBiomass Digester 14".

AUTHOR INDEX